Internet of Things Security
Challenges, Advances, and Analytics

Internet of Things Security

Challenges, Advances, and Analytics

Patel Chintan
Nishant Doshi

CRC Press
Taylor & Francis Group
Boca Raton London New York

CRC Press is an imprint of the
Taylor & Francis Group, an **informa** business

CRC Press
Taylor & Francis Group
6000 Broken Sound Parkway NW, Suite 300
Boca Raton, FL 33487-2742

Printed on acid-free paper
Version Date: 20180723

International Standard Book Number-13: 978-1-138-31863-2 (Hardback)

Visit the Taylor & Francis Web site at
http://www.taylorandfrancis.com

and the CRC Press Web site at
http://www.crcpress.com

To my Parents, Wife, Child, Guide, Friends, Colleagues, and God.
- Chintan Patel

To my Parents, Wife, Guide, Friends, Colleagues, and God.
- Nishant Doshi

Contents

Foreword

Internet of things(IoT) and its applications are quickly receiving the attention of enterprises, business communities, research communities and government. Governments from both developed and developing nations want to adopt the internet of things and want to re(design) their cities in such a way that a manually functioning city will be converted into a smart city. A smart city will have certain other applications like:

- Smart home

- Smart grid

- Smart gas, water distribution

- Smart parking and traffic system

- Smart environment monitoring

- And so on......

So during the journey of these technological as well as societal developments, contributions from the students and research community will play a significant role. This book will provide an opportunity to every student, researchers and general people to learn about basics of the internet of things. Chapter 1 provides in-depth knowledge about the internet of things and its vision with various applications. As the title suggests, the motive for writing this book is to discuss the various security challenges residing in the internet of things, Chapter 2 to Chapter 7 discuss the basics of security to advanced security analysis tools.

Basic knowledge of mathematics is the key requirement to understand cryptography and security. Chapter 2 provides a clear understanding about cryptographic maths in very simple and generalized terms so that any newcomers can also learn it. Chapter 3 discusses topics regarding authentication in cryptography. Successful authentication is the key challenge in the internet of things. This chapter will provide the in-depth understanding of various scenarios and phases of authentication.

Chapter 4 and Chapter 5 provide detailed authentication flow in the single server and multi-server environment. So any researcher who is focusing on design and development of authentication scheme can have detailed knowledge about phases of authentication in both single and multi-server environment. Chapter 6 discusses various possible attacks and scenarios, which create vulnerability inside the system. A most beneficial aspect of this chapter is that it provides remedies for the attack also so researchers can have a clear guideline that they need to follow to make a secure design from the various attacks.

Chapter 7 discusses various analytical matrices to compare the designed security algorithm with others in terms of time complexity, space complexity, and energy consumption.

Overall, I believe this book will surely prove to be starting point from stepping stone to reaching the peak of the mountain for the internet of things lovers. So welcome to the journey of the internet of things and security of it. Turn the pages and make a deep dive into the world of knowledge. All the best.

Dr. Tajinder Pal Singh
Director, School of Technology
Pandit deendayal petroleum university, Gandhinagar

Preface

The recent advancement in technology has opened the door to visualize the technological world in the new direction and new hope. In 1980s the internet was big revolution, and it had captured significant growth in 1990s and 2000s. Internet of things, initially, was less focused due to certain reasons like less availability of literature and low industrial acceptance. The advancement of sensing technology attracted the researcher community to work in the internet of things. Advanced sensing technology has drastically updated data collection methodology. Internet of things applications like smart home, smart health care and industry 4.0 had grown/ will grow significantly between 2010 - 2025. Prediction of more than 75 billion devices by 2030 will surely become true if current growth of devices will continue for next 10 years. Any technology comes with both hope and challenges. Internet of things came with the hope of better life, better services, quality manufacturing and so on. With this hope internet of things came up with various challenges like reliable communication between devices, collision free communication, handling mobility of devices and users and so on. In this book, we have focused on the challenge called "**Security**". Cyber attacker community has contributed lot in the development of cyber security and cryptography. Intelligent cyber attackers have forced the researcher community to design a secure ecosystem for the internet. Various algorithms and protocols are developed for encryption/decryption, hashing operations and other cryptographic requirements. Required combination of encryption algorithm and hashing algorithm has achieved required security level. But....

Internet of things communication is different than internet. Major challenges in the internet of things, which are absent in the internet of things can be listed as below:

- Resource constrained devices.

- Absence of globally accepted standardization.

- Heterogeneity of protocols.

In the internet of things, TCP/IP model is globally accepted architecture derived from globally accepted OSI model but in internet of things, there is no availability of globally accepted standardization. Every industry is developing IoT devices with their own visualization and research outcomes. When the situation comes and every differently manufactured device needs the communication then it will be a very difficult task to enable communication. Another major challenge is lower resource availability with devices. Resources can be considered as a battery and memory. So to tackle with security algorithms, we need to be lightweight in terms of computation and resource requirement so we can implement those algorithms in tiny devices of IoT also. Cryptography has three major pillars,

■ Confidentiality.

■ Integrity.

■ Availability.

These three pillars can be achieved by implementation of secure **"Authentication"** scheme. So in this book, we have discussed the internet of things and its authentication scenarios. We have observed need of secure authentication scheme to enable secure communication between devices. Authentication provides trust to the communicating entities that they are communicating with valid devices in the network and also communicating in a secure manner. So major chapters of this books are written in very simple manner so any new researcher who wants to work on authentication can understand the scenarios of authentication and contribute in the advancement of IoT security.

Chapter 1 titled "IoT: An overview" discusses introduction to internet of things, which will provide deep and clear understanding about internet of things, internet of things probable reference architecture and internet of things security. In this chapter, we have also discussed light weight cryptography and the needs for it.

Chapter 2 titled "A mathematical foundations" provides a clear understanding about various cryptographic algorithms and mathematics used. We have tried to start with basic level of mathematics and reached to advanced level of mathematics used in authentications and IoT security. This chapter also highlights the elliptic curve based cryptography and mathematics behind it. Overall this chapter will make clear mathematical fundamentals for readers.

Chapter 3 titled "IoT authentication" discusses various authentication scenarios in the internet of things. Authentication layered architecture discusses opportunities and scenarios, which will create the need for secured authentication protocols in the internet of things. Authentication phases are discussed with graphical representations in this chapter, so every reader can easily understand the communication structure in authentication. The cloud centric authentication topic provides clear understanding about end-to-end authentication scenarios for fool-proof secure end-to-end communication. Application oriented authentication scenarios discussed in the chapter will enlighten readers about possible points where authentication will be required during development of application.

Chapter 4 titled "Single server authentication" opens the door to understanding, how to design secure authentication protocol in the environment of single server. Single server can be fog device or gateway also. Designing peer-to-peer secure authentication scheme to produce a design complete secure system is the main focus of this chapter.

Chapter 5 titled "Multi server authentication" focuses on designing an authentication scheme for the environment where users want to communicate multiple servers and need to register with multiple servers. Authentication scenarios discussed in this chapter will provide deeper knowledge about how make use of registration center and how to design the scheme where only-one time registration will enable communication with multiple servers.

Chapter 6 titled "Attacks and remedies" discusses various attacks possible in the developed authentication schemes. We have provided figured examples of attacks, so every user can easily understand that in these circumstances, this type of attack is possible. This chapter also shows remedies for these attacks so new authentication scheme designers can keep that in mind and design schemes in such a way so that they will not prone to these attacks.

Chapter 7 titled "Analytical metrics and tools" provides detailed information about parameter, which makes any security protocols and algorithms better. Time complexity, space complexity, and energy consumption will be the most significant parameters for designing and developing secure schemes. This chapter also discusses about various analytical tools, which can be used to analyze security protocols designed by the researcher. These tools provide knowledge about self evolution of security schemes. In this chapter four major tools are discussed: AVISPA, BANLogics, Scyther and ProVerif with the example of Needham Schroeder protocol.

Chapter 8 titled "Future works and conclusions" is a small chapter that discusses future work that can be carried out to protect the IoT ecosystem from cyber attackers and concludes the complete book.

Acknowledgment

I would like to thank many people and want to show my heartfelt gratitude towards them. Without their endless support, it might not be possible to put this IoT authentication text book in front of people.

I would like to thank Taylor & Francis group and CRC press from the bottom of my heart to accept my proposal for publication of this book. I would like to show my gratitude to Rich O'Hanley for support and guidance regarding book writing and designing.

I would like to thank my native institute Pandit Deendayal Petroleum University, Gandhinagar, Gujarat, India and its management for their support in terms of motivation and resources. I would like to give my heartfelt thank to university director general Shree Dr. C. Gopalkrishnan and director, school of technology, Shree Dr. T. P. Singh sir for his continuous guidance in research skill development and motivation for publication with world class publications.

I would like thanks to my co-author and my PhD guide, Dr. Nishant Doshi for the continuous motivation. Without your guidance and editing activity it would not be possible to work on this book.

I would like to thank my mother and my father, without your blessing it was impossible to persist in this long journey and also to my doctor sister, my highly motivated wife, my lovely son and my rest of the family for their continuous support and motivations.

I would like to thank my Ph.D friends, especially who always supported and motivated me in the shaping of this wonderful book.

Last but not the least, thanks to God who was ultimate source of energy to do so. Thank you all ....

-Chintan Patel

Authors

Mr. Chintan Patel
Pandit Deendayal Petroleum Univ.
Gujarat, India
https://orcid.org/0000-0002-3824-6781

Dr. Nishant Doshi
Pandit Deendayal Petroleum Univ.
Gujarat, India
https://orcid.org/0000-0003-3443-7561

Symbol Description

$\mathbb{N}$	Set of all the natural numbers	$\triangle$	Binary operation
$\mathbb{I}$	Set of all the Integer numbers	$\square$	Unary operations
$\mathbb{Q}$	Set of all the rational numbers	$E(X,Y)$	Elliptic curve on variable X,Y
$\mathbb{P}$	Set of all the prime numbers	∇	Slope of an elliptic curve
$\mathbb{R}$	Set of all the real numbers		
$\mathbb{C}$	Set of all the complex numbers	$e: X \to Y$	Bilinear Pairing
$\mathbb{G}$	Set of all the group members	$\mathbb{Z}_i *$	Set of prime integers
$|\mathbb{G}|$	Order of the group	$\mathbb{GF}$	Finite field
$\sum$	Sum of all the terms.	ϕ	Euler's totient function

Chapter 1

IoT: An Overview

CONTENTS

√ If you think that the internet has changed your life, think again.
The Internet of Things is about to change it all over again!

Mr. Brendan O Brien
Chief Architect & Co-founder of Aria Systems

1.1 Abstract

Internet of things has created significant impact for the technological researchers. Different research communities have started to explore different domains, different technologies and different applications for the internet of things. In this chapter we have introduced internet of things, internet of things vision, basic architecture of the internet of things, IoT security, and lightweight cryptography. Overall this chapter will provide a basic understanding about internet of things and its nuts and bolts.

1.2 Introduction

Kevin Ashton, A founder of Auto-ID Center at MIT sparked a word "Internet of Things (IoT)" in 1999 during a proctor and Gamble presentation[Ashton (2009)]. In a simplified way, IoT is a bridge from physical world to cyber world. In the modern wireless communication world, IoT is the paradigm that is gaining tremendous focus of researchers, industries and governments. The term *Thing* in the IoT refers to a device having following properties[Al-fuqaha et al. (2015)]:

■ Unique identity

■ Capable to connect with other devices

■ Self powered with long lasting battery

■ Capable to perform control commands

Figure 1.1 shows basic architecture of internet of things. U.S. national intelligence council included IoT as one of the *disruptive civil technologies* with potential impacts on U.S. national power.

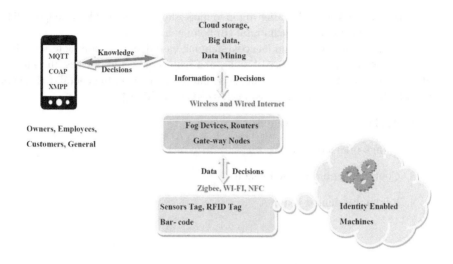

Figure 1.1: IoT communication structure.

As illustrated in Figure 1.1, the base level in IoT consists of identity enabled smart machines. These smart machines can be sensors , actuators , devices capable to sense, read, hear, think, see and communicate. These machines will be connected with fog devices or gateways that work as a interface between physical objects and cyber world. Fog devices [You and Learn (2015)] are smart routers that are capable of doing some data processing. Fog devices will convert the data into the information and pass this information to the cloud . In the IoT, collection of instructions will be in three basic forms as follows[Borgia (2014)] :

■ *Data* : Generated by sensors, larger in size, completely unstructured

■ *Information*: Generated by gateways and fog devices, medium in size, little structured

■ *Knowledge*: Generated by data miners at cloud, very small in size, fully structured

Cloud receives the information from the various gateways and fog devices. Size of this information is dependent on application and number of devices implanted at base level. Data miners at cloud will do the mining of this information and generate knowledge from them. This knowledge can be in the form of control messages, instruction and alert. Only users who are allowed to access this information via applications will get access to this knowledge after log in to the system. Generated knowledge will be completely structured and meaningful [Bonomi et al. (2012)].

Complexity of the IoT and its ecosystem increase due to several factors not limited to increases in the number of smart and battery-powered devices, the number of users, distance between communicating entities and so on. In general, IoT can be broadly divided into four basic visions [Atzori et al. (2010)].

1. Application-oriented vision

2. Things-oriented vision

3. Communication-oriented vision

4. Research-oriented vision

1.3 IoT Vision

Vision for the IoT generalizes a futuristic aspect of world. A common vision is "Smart world".

1.3.1 Application-Oriented

A successful implementation of IoT will impact the economy of world greatly. IoT provides tremendous opportunity to the device manufacturers, application developers and internet service providers. Business of smart objects of IoT is predicted to reach *2 trillion* models to be deployed by 2020[Al-fuqaha et al. (2015)]. By 2022, 45% of world internet traffic will flow from machine to machine communications. McKinney's report shows that by 2020, 30 billion devices will be connected with each other. McKinney's report on global economy estimate that IoT will have a total potential economic impact of $39 trillion to the $11 trillion by 2025[Al-fuqaha et al. (2015)]. Figure 1.2 shows basic application-oriented vision for internet of things. Impact of IoT on manufacturing can be better understood by calling it industry 4.0. Industry 4.0 is a combination of five enabling technologies. This includes IoT, Cyber Physical System[McKay et al. (2017)], M2M communication[Borgia (2014)], cognitive computing[Wu et al. (2014)] and big data[Sun et al. (2016)].

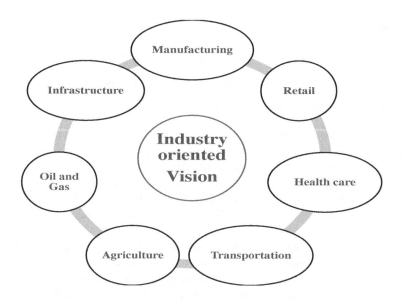

Figure 1.2: Application-oriented vision.

According to data released by international data center, $178 billion was invested in IoT based manufacturing in 2016-17. "Sufficient automation and efficient optimization " are major drivers for the manufacturing industry[Xu et al. (2014)]. Manufacturing involves two main communication environments: human-to-machine and machine-to-machine. Important features of smart manufacturing applications involves :

■ Large network of sensors and RFID based machines

■ Dynamic ecosystem and topology

■ Cloud computing and big data

■ Machine to Machine and Human to Machine

■ Advanced ecosystem

IoT based industrial plant monitoring will reduce the number of accidents and provide the better security in dangerous plants like gas based, nuclear based, oil based, petro chemical plants and so on. Each of these applications requires use of various sensors to do several tasks like monitor the pressure, temperature, speed, location. Thus, the sensors who operated at base level play a crucial role in the security of these plants[Xu et al. (2014)].

Retail and logistic industries are one the fastest growing industries and they are adopting technology to provide efficient services to the customer. In the retail and logistic industries IoT can be used for the following purposes:

■ To track the inventory

■ To manage the warehouse efficiently and handle retail management

■ To provide accurate knowledge about quality and quantity of the inventory

■ To track the life cycle of objects

Retail industries use radio frequency identification(RFID)[Dusart and Traor (Dusart and Traor)] technology for the above mentioned applications. Functionality like monitoring of production process to disposal will be implemented by RFID tag and RFID readers. Adoption of IoT reduces material waste, lowering cost and improving profit margin for both retail and logistic industries. For the food and liquids logistic, IoT will help to ensure the quality of the product in terms of freshness of perishables dates.

Temperature sensors and humidity sensors can be used to monitor continuously temperature and humidity of the products during transportation respectively. Actuators fitted in the containers will maintain the temperature inside the containers based on requirements of materials. Another concept is smart mall in which the customers, shopping habits and item choices are analyzed. Such systems will be used to recommend shopping destinations, discounted products, ATM and payment options to the customers.

Adoption of IoT in social life development is very much essential. Health care [He and Zeadally (2015)] is one of the fastest growing social applications and also a industry that is gaining benefits of technology advancements. For health care, IoT can be used for :

■ To monitor health services of the individual as well as the city

■ To monitor health inventory

■ To provide on time health service to the needy

Various vital functions like temperature, blood pressure, cholesterol level, heart rate and so on can be tracked by various medical sensors. Smart beds in the hospital rooms will take care of patient foods, entertainment, medicine. Smart rooms of the hospital will be developed in such a way that if the patient is going to take more time in hospital, then it can also take care of education using smart classes and virtual classes.

Adoption of technology in agriculture[Abouzahir et al. (2017)] and breeding will increase the income of farmers, develop the quality of food, provide better services to animals. Smart veterinary is the concept that can help the local municipal bodies keep track of the health of animals. Advanced sensors may capture the information about infections to the animals and inform the authorities. It can help in reducing the spread of infection to the other animals. The temperature of the animal is a very important parameter for the animal health. Changes in the temperature of any animal can help the doctor to easily identify the diseases. Governments can use the collected data to announce various schemes for the betterment of animal health.

Monitoring the plants and lands by using advanced microchips is the ongoing development of smart agriculture. Plants can be monitored by the sensor that can measure the additives, melanin and fertilizers. Smart agriculture systems can be viewed in two ways: one is in terms of plant monitoring and other land monitoring. Land monitoring can be done by using moisture sensors, temperature sensors and so on. Smart agriculture[Abouzahir et al. (2017)] will lead the government to develop real time tracking of food productions, lands, registration and monitoring, fertilizer requirements of individual farmers and so on. Farmers can start their own *E-Mandi* through which they can sell their products directly to the customers. Smart agriculture includes smart irrigation systems that will study the moisture level of land using moisture sensors to provide water to the plants. Another major domain of IoT is smart city. Various sub domains of smart city include smart grid, smart home, smart transportation, smart tourism and smart environments.

Smart home [Yassein et al. (2016)] means making devices smarter. Various devices like TVs, mobiles, laptops, speakers, lights, window shades, AC, meters, clocks will become smart devices so each and every device will take care of moods, choices and feelings of a person and behaviour. Device area networks will allow devices to communicate with each other. Lights will constantly monitor intensity of sunlight and change the intensity of light based on that. Smart air conditioners will monitor the temperature of the home continuously and based on that, automatically maintain the room temperature. A person's car will communicate with the lights, AC, and coffee maker so that it can take care of the comfort of the person. Home automation systems will enable controlling the complete home from anywhere. Indeed, IoT is services from *anywhere, anytime, any place.*

Electricity controlling can be another major advancement using IoT. Maintaining the electricity cycle in the home will lead towards reduction of bills and electricity usage during the peak times. Advanced level monitoring of data captured from smart technologies will be input for cloud computing, data mining, big data, machine learning and neural networks.

Smart grid [Aloul et al. (2012)] is defined as a intelligent electricity distribution system that can keep track of electricity flow, health of the grid, controlling flow of energy, and storing the energy. A complete smart grid can be combinations of smart controlling devices, smart meters, smart home electric appliances, smart switches. Currently energy distribution is one way (from producer to consumer). Another concept which is discussed about the smart grid is bidirectional energy distribution in which consumers may also generate energy using various renewable energy sources like solar, water and wind. A solar farm is a place in which farmers generate the energy by solar plates or wind turbines. Later on, they can sell this energy to the government and other consumers. Major requirement of the smart grid is proper electricity distribution, taking care of failures during rainy days or during natural or human disasters. Smart grid connected with smart transportation will help in faster recharging of e-vehicles. Currently in the e-vehicles, the major challenge is the recharging of battery during long-distance driving. Smart identity cards can be used for the identification of consumers and provide better charging services. Vehicles to grid communication may enable vehicles to communicate directly with the electricity grid for recharging systems. Smart grids can be used to monitor the energy usage habits of consumers, homes, cities. Service provides can analyze the gathered data so that they can identify the peak time for high energy requirements.

Smart transportation [Chaturvedi and Srivastava (2017)] includes smart roads, smart vehicles, smart tolls, smart plants, smart parking, smart fuel stations, smart ambulances and smart environments. Smart devices in the vehicles can collect the data from the roads about road traffic intensity, about accidents, and can gather the information about free parking slots from smart parking. Government authorities can monitor the parking situations in the city using various cameras and motion sensors. Smart tolls can be an easy payment system for toll tax on roads and will reduce the fuel consumption as well as transportation timing. Near field communication enabled vehicles or phones will pay for parking, pay for fuel as well as pay for toll tax. RFID systems can also be used to add advance features in vehicles.

IoT can be used to improve the life of disabled dependents by improving social inclusions, elderly assistance, and disabled assistance concepts. Smart buildings can have various applications like plant monitoring to keep track of oxygen status in the building, energy maintenance, child protection and video surveillance.

IoT applications include mobile application development, authentication and access control for the devices and humans, cloud computing concepts, big data gathering and analysis, machine learning, and many more advanced technologies.

1.3.2 Things Oriented

"Things" in the IoT are the physical devices or virtual devices that are involved in the successful implementation of smart system"[Ashton (2009)]. IoT can be seen as a dynamic global network infrastructure with self capabilities based on standard and inter-operable communication protocols where physical and virtual things have an identity, physical attributes, virtual personalities, user intelligent interfaces, and are seamlessly integrated into the information network. Things oriented vision is divided in four major parts :

(A) **Collection / sensing part:** In this part, devices mainly focus on data collection and data gathering functionality. These functionalities can be achieved by using cameras, sensors, voice recorders.

(B) **Local processing and transmission part:** In this part, devices will focus on local processing of the data and transmission of the data. These functionalities can be achieved by using fog devices and smart gateways/routers.

(C) **Information processing and knowledge generation part:** In this part, devices will focus on high level data mining and knowledge generation. These functionalities can be achieved by using cloud infrastructure either virtual or physical.

(D) **Controlling part:** In this part,devices will focus on controlling these complete applications and it can be achieved by using either mobile applications or computers.

Here we will focus on physical things. Data collection or data gathering can be handled by using various sensing devices with smart sensors, smart cameras, smart satellites[Summary, Latency, Rate, and Latencies (Summary et al.)] .

Wireless sensor network is the collection of various battery-powered devices that can sense the environment and send the data to the local gateway. Wireless sensor networks are used for many different domains of research ranging from agriculture monitoring to underwater sensor networks. Small batteries and lower memory of the sensors are two important parameters that have attracted the focus of researchers. Traditional wireless sensor networks consists of static and resource constrained sensing devices in the area.

Wireless Medical Sensor Networks [Ferrag et al. (2017)] (WMSN) is the collection of various medical sensors that are placed on doctors, patients, nurses, hospitals. As we discussed, early identification of disease can save many lives. Major sensors used in the medical industries are temperature sensors, blood glucose sensors, blood oxygen sensors, ECG Sensors, image sensors, internal sensors, pressure sensors, and motion sensors. There are many other sensors used in WMSN such as strip sensors, wearable sensors, implantable sensors, invasive sensors, non-invasive sensors and indigestible sensors.

Wireless agriculture and forestry sensor networks [Abouzahir et al. (2017)] are focusing on enhancement of services to the animals and high quality and quantity of food production. Sensors used in this network are temperature sensors, moisture sensors, humidity sensors, motion sensors, sensor cameras, pressure sensors, and fire sensors. Early fire detection in the forest can save the many animals and many trees. Sensors used for chemical analysis maps, soil nutrients, pesticides, herbicides and fertilizers. Other sensors used in farming provides navigation, crop row detection, plant alignments, obstacle avoidance, crop rows detection, soil compaction, canopy analysis, 3D structures, and bark thickness. Location sensors also play major roles in the land monitoring. Sensors connected with satellites will continuously transmit the data about latitude and longitude of the land. So smart phone applications for agriculture include disease detection, soil and water study, crop harvest readiness, and fertilizer calculators. Building mobile sensors like accelerometers can help to determine leaf angle index or gyroscope can be used for detecting equipment rollovers. Mobility management in the wireless sensor network enables mobile device to collect the data whenever the opportunity rises. Most of the sensors in the wireless sensor network do the communication at 2.4 GHz band with the 250 kb/s.

RFID Tag is one of the most important things used for data acquisitions. RFID tags enable advanced capabilities in various objects like human, animals, any other objects. The most important capability is unique identity. Using RFID we can easily do the identification of objects, we can track the object, and we can monitor the object. RFID enables wireless communication in each object in such a way that it can transmit attributes of objects, and identity of objects to the other electronic devices. RFID tags can be mainly of two types: active tags and passive tags. Most important difference between these two tags are battery supply and the distance covered. Active tags have their own power supply. They focus on regeneration of energy by generated energy. Active tags cover longer distances and are able to perform high level difficult operations. Passive tags do not have power supply, but they transmit data by using the energy that the reader generates during its functionalities. Passive tags are smaller in size, less costly, with longer life and very short range (approximately 3 meters). Every object used in the RFID based system will be equipped with RFID tags, and these RFID tags have one code called an electronic product code. RFID readers will collect the data from the RFID tag and will transmit that data to fog devices for more processing and transmission in the internet [He and Zeadally (2015)]. There are four frequency ranges of communication between RFID Tags and RFID readers.

1. **Lower frequency:** 125 kHz/134 kHz and 140 kHz/148.5 kHz.

2. **High frequency:** 13.56 MHz.

3. **Ultra high frequency:** 915 MHz in US and 868 MHz in Europe.

4. **Microwave:** 2.4 GHz and more.

Selection of frequency ranges depends on applications. RFID follows ISO/IEC 18000 standards, most of the sensors follows IEEE 802.15.4, ZIgBee[C, Muthu Ramya, Shanmugaraj.M (2011)] , Wireless HART[Chen et al. (2014)], ISA100 standards for the communication.

Fog devices are the routers used for data forwarding as well as data processing. Fog devices are also called smart routers that are capable of transmitting the data on the internet at the same time it processes the row data generated by the sensors.

As shown in Figure 1.3, sensor nodes are the devices that can gather the information from the environment and communicate with other sensor nodes. All sensor nodes have following components :

1. Data capturing hardware

2. Power supply

3. Storage

4. Controller to control and data processing

5. A transceiver for data and frequency transmission

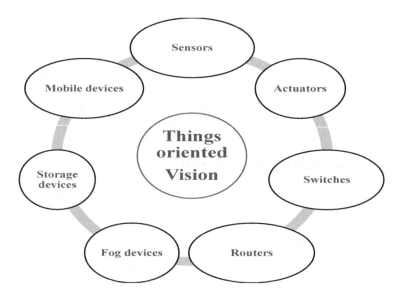

Figure 1.3: Things oriented vision.

Fog computing is another important advancement of networking and fog devices, and it plays a major role in the successful implementation of it. Fog computing adds distributed computing (simple means like distribution of computing) in the IoT. The IoT era before fog computing was as follow:

1. Sensors at ground level collect the data and transfer to the cloud via routers or gateways

2. Big data generated in the cloud will be processed and based on that knowledge will be generated. [All data processing in cloud]

IoT era after fog computing is as follows :

1. Sensors at ground level collect the data and send that data to the fog devices /intelligent routers

2. Fog devices will do the processing of this data and will transmit this generated information to the cloud

3. Cloud will do the processing of information (processed data) and generate knowledge

Fog devices also provide great mobility, high scalability, low latency and heterogeneity. So the basic goal of fog computing is to reduce the data volume and traffic to the cloud server, decrease the latency and improve the quality of service. Major security challenges in IoT fog devices are authentication and trust of valid fog devices, detecting malicious nodes, privacy and access control, intrusion detection, and data protection.

Major players in development of IoT based fog devices are Samsung, Smart-Things IoT hub, Alphabet's Nest, Amazon, Google, Intel, Microsoft. Fog devices are also used for device authentication and identification. Fog devices validate the certificate for the devices connected. So fog computing involves data processing, security analysis and data forwarding [Yi, Qin, and Li (Yi et al.)].

1.3.3   Communication Oriented

Communication in the IoT will be required to take care at following situations:

1. Communication between sensor devices and actuators [perception layer]

2. Communication between sensor devices/actuators and fog devices /gateway

3. Communication between fog devices and cloud

4. Communication between user devices [mobiles] and cloud

5. Communication inside cloud platforms

As shown in Figure 1.4, we have divided communication protocols in the IoT into mainly two parts: short-range protocols which are used in communication between sensor devices and between sensor devices and fog devices while long-range protocols are used for communication between fog devices and cloud as well as between user and cloud services.

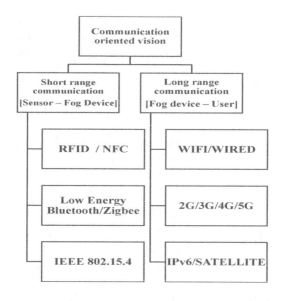

Figure 1.4: Communication oriented vision.

Protocols which are part of short-range communications are wifi[Varma (2012)], ZigBee [C, Muthu Ramya, Shanmugaraj.M (2011)] , z-wave [Yassein et al. (2016)], RFID,NFC [Communication (2011)], Bluetooth Low energy [Mackensen et al. (2012)], Thread, Home plug and DSRC [Kenney (2011)](Dedicated short range communication). Protocols which are part of long range communications are Sig-fox [Lauridsen et al. (2017)], Neul, LoRaWan [Al-fuqaha et al. (2015)], 6LoWPan [N. Kushalnagar (N. Kushalnagar)], GSM/3G/4G/5G and WiMax [Max and Etemad (2009)].

Table 1.1 discusses basic communication protocols and technologies that can be useful in IoT implementations. Availability of protocols like bluetooth, ZigBee, Wireless-HART [Highway addressable remote transducer protocol] enable development of communication enabled tiny sensors. IEEE 802.15.4 standards provide development of low data rate, low power small area networks. Wireless-Hart was developed by the HART communication foundation for the requirement of processing wireless field area networks.

Table 1.1: IoT Communication technologies.

Range	Technology	Standard	Frequency	Bit rate	Coverage
short	WiFi	IEEE 802.11	2.4/5 GHz	500Mbps	50 m
	Zigbee	IEEE 802.15.4	24 GHz	250kbps	100m
	Z-Wave	ZAD12837	900 MHz ISM	40 kbps	50 m
	RFID	ISO/IEC 18000	LF,ISM Band	40 kbps	<2 m
	NFC	ISO/IEC 18092	13.56 MHz	424 kbps	<20 cm
	Bluetooth LE	IEEE 802.1	2.4 GHz	1 Mbps	50 m
	Thread	IEEE 802.15.4	2.4 GHz	250 kbps	<100 m
	Home Plug	IEEE1901	<100 MHz	10-500 Mbps	<100 m
	DSRC	IEEE 802.15P	5 GHz ISM	4Mbps	300 m
Long	Neul	Neul	458 MHz	10 kbps	10 km
	SigFox	Sigfox	900 MHz ISM	1 kbps	10 km
	6LoWPAN	RFC6282	ISM Band	N/A	N/A
	GSM/ 3G/ 4G/ 5G	GSM, UMTS, HSBA, LTE	900/ 1800/ 1900/ 2100 MHz	10 Mbps	50 km
	WiMax	IEEE 802.16	2.3, 2.5 and 3.5 GHz	410 Mbps	510 km

Bluetooth low energy is a highly effective communication technology where the battery saving is an important aspect. It has low power consumption and good data throughput. It supports 8-bit micro controllers. BLE consumes energy for connection setup up to 0.3 mJ. BLE consumes energy for a data transmission with a notification packet up to 0.13 mJ. Energy consumption for single data transmission if the transceiver is switched on/off between the data transmission is up to 0.78 mJ. Stack size for BLE slave devices is 82KB of ROM and 1 KB of RAM, while master device uses 70KB of ROM and 1KB of RAM.

Home-plug organization [Latchman et al. (2013)] developed Home-plug 1.0 standard for the power line communication system. Home-plug AV (audio-video) provides data, video, audio communication using power line communication. Home-plug 1.0 provides peak PHY Data rate 14 Mb/second. Medium access control layer of Home-plug uses career sense multiple access as a protocol for channel accessing. Home-plug 1.0 does not provides any encryption and security standard. Home-plug AV provides key distribution and also uses AES 128 bit encryption standards.

WiFi (wireless fidelity) [Varma (2012)] is a IEEE 802.11 wireless local area network. It is a subset of IEEE 802 set. Its collection of Wireless LAN and Wireless MAN.802.11 supports 1 or 2 Mbps transmission in the 2.4 GHz band and uses frequency hopping spread spectrum (FHSS) or direct sequence spread spectrum (DSSS). 802.11b, an extension to the previous one provides 11 Mbps transmission (with a fallback to 5.5, 2 and 1 Mbps) in the 2.4 GHz band. 802.11b uses only DSSS. 802.11a provides communication up to 54 Mbps in the 5GHz band. 802.11a uses an orthogonal frequency division multiplexing (OFDM) encoding scheme rather than FHSS or DSSS. 802.11g is an extension to 802.11b that provides up to 54 Mbps in the 2.4 GHz band. 802.11g also uses OFDM.

ZigBee [C, Muthu Ramya, Shanmugaraj.M (2011)] provides reliable, low cost, long battery life and remotely upgradable firmware. It was developed by Zigbee alliance. It is a parallel standard with IEEE 802.15.4. It provides range up to 150 meters outdoors using a method called direct sequence spread spectrum (DSSS). DSSS consumes less power compared to other spectrums like frequency hopping spread spectrum. Zigbee has low throughput and a data rate upto 250 Kbps. ZigBee system is useful for applications where low data rate is required. Example of such applications are home automation and control, smart meter reading, smart building and residential systems. E-health and body area networks. ZigBee based smart energy, hospital & institutional, patient monitoring, cable replacements, automotive, in vehicle control, it is used for vehicular, entertainment, status monitoring, telecommute services.

WiMAX(Worldwide inter-operability for microwave access) is the group of standards on IEEE 802.16. WiMAX was developed with the aim of providing 30 to 40 Mbps; currently it supports 1 Gbps. WiMax can be used to provide portable mobile broadband connectivity in the big cities. It is also capable of voice over internet protocol. Subscriber stations are the devices that are used for the connection with WiMAX. WiMAX supports 2.3 GHz, 2.5 GHz and 3.5 GHz spectrum for the communication. WiFi provides connectivity with in 300 feet, while the WiMAX provides connectivity up to 30 Miles.

RFC 4919 [N. Kushalnagar (N. Kushalnagar)] defines IPv6 over low-power wireless personal area networks. A LoWPAN is a low-cost communication network that connects real time physical environment devices like wireless sensors. It has small packet size. At the physical layer, its package size is 127 bytes. It supports 16 bit or 64 bit media access control addresses. It supports 250 kbps, 40kbps and 20 kbps data rates for 2.4 GHz, 915 MHz and 868 MHz physical layer. It provides low power and low cost power consumption. Reliability of communication is low compared to other protocols. It provides built-in security mechanism using AES and SHA.

Radio frequency identification followed quick response code for the device identification. RFID is composed of mainly two parts: one is RFID tag and other is RFID reader. RFIDs have some good plus points compared to QR Codes. One of the most important advantages is its capability of communication without physical connection or in front connection. RFIDs have two types of tag: First is the active tag that requires a battery as a power source. Battery cost, size and life time are not suitable for some real time scenarios where mobility of objects is very high. A passive tag has three major part: an antenna, a semi conductor and encapsulation. Tag reader is responsible to power tag. Tag antenna will collect the energy and transfer the tag's id. Encapsulation works as security for the RFID. RFIDs can be built with various sensors. So it can work as identification for the objects and sensor based functionality implementations. RFIDs require small storage size of 200 to 8000 bits. RFID works on low frequency, high frequency and ultra high frequency. Near field communication works in high frequency bands at 13.56 Mhz under ISO 14443 and ISO 18092. NFC supports maximum data rates of 424 Kb/s up to 10 cm. NFC enables communication between active reader and active tag with peer-to-peer communication. NFC follows NFC data exchange format message for the data transmission. NFC is one of the most prominent technologies for upcoming smart phones, which are not just the smart phone but also a wallet.

In 1994, when Ericsson mobile communication started to work on low power consumption system [From (2005)] to replace cables for the short range communications, it came up with the technology called Bluetooth. Bluetooth is short range communication device suitable for mobiles, mike, keyboards and many more. Bluetooth devices can communicate either in master mode or slave mode. It supports at max eight devices to connect in piconet. Bluetooth uses 2.4 GHz band, which is not licensed in any country and available free of cost. Bluetooth supports a synchronous connection less links and synchronous connection oriented links. Bluetooth devices have 48 bit unique identity. Bluetooth provides security in three basic modes.

1. Not secure

2. Service level security

3. Link layer security.

Bluetooth was developed for portable products and offers short range. Bluetooth uses very low battery. Wi-Fi was developed for longer range and consumes high battery. Bluetooth devices consume 1-35 ma while Wi-Fi devices consume 100-350 ma. Bluetooth handles power in two modes: one is stand by and other connection. Wi-Fi have an active mode and a power save mode. Bluetooth uses shared secret and pairing for authentication, while Wi-Fi uses shared secret and challenge response for authentication. Bluetooth makes use of E0 stream cipher, and Wi-Fi uses RC4 stream cipher for encryption. Z-Wave is wireless protocol used in home automation. A smart home network enabled with Z-Wave contains 232 appliances. Appliances can be either controllers or slaves. Z-Wave works on MAC Layer, Transport layer, Application layer and routing layer. Comparatively Zigbee provides better reliability, low radio rebirth, easy usage and easy inter-operability [From (2005)].

Dedicated short range communication technology was developed by the automotive industry to develop a smart transport system. Wireless access in vehicular environment uses, the IEEE 1609.2, 1609.3 and 1609.4 standards for security. DSRC is currently in an under development phase. Basic motive behind deployment of DSRC was to make collision prevention techniques that depends on data exchange in vehicular network. At MAC and PHY layer DSRC utilizes WAVE protocol. At network layer, it uses 1609.4 for channel switching, 1609.3 for network services, and 1609.2 for security services.

LORA [Lauridsen et al. (2017)] is another type of LPWAN that provides long-range low-power wireless connectivity for sensor devices, a major application of LORA is smart electricity meters. Wi-SUN is used for connectivity in home devices. Narrow Band IoT and LTE-M can be used to connect millions of sensors and devices using cellular technology.

There are many other protocols like RPMA [Landscape and Technology (Landscape and Technology)] and many companies that are working on development of standards for IoT communications. It keeps various features in mind during this development.

Features of IoT Communications

■ Ground level networks will be heterogeneous; they will be a combination of wired, wireless and cellular networks as well it support different communication modes either point to point or client server.

■ Support any type of addressing modes; it should support unicast, multicast and broadcast. Some time enhanced broadcast service to unicast and multicast.

■ Efficient medium access protocol.

■ Guarantee of reliable communication.

■ Better access control and strong scheduling algorithms.

■ Strong and smooth mobility model.

■ Correct, short, congestion free and optimal paths.

■ Less computation, Low power consumption and better services.

■ Alerts and notifications.

■ Security, proper device authentication, data access, confidentiality and integrity.

■ Proper interface design.

Congestion free, reliable communication is the fundamental aspects for any internet-based technology so it becomes important to focus on it. Another vision for the IoT is a research based vision in which we will basically discuss various areas on which researchers are focusing.

1.3.4 Research-Oriented

There are many contributors in the research-oriented vision. First and most important contributors are R&D departments of various industries. Other contributors are researchers working in academics, government and organizations. Similarly many other entities are involved in the successful deployment of any system. We will discuss our research oriented vision as per following figure.

Figure 1.5: Research oriented vision.

As shown in Figure 1.5, there are seven types of vision that need to focus on IoT.

■ **Sensor as a Service**:

Sensor is a very small but most important device in the IoT. Sensors with the identification tags like RFID or NFC can play a role as an object identifier and attribute identifier. Temperature sensor set with RFID tag on a container can help to maintain the temperature of the container as well as identify the objects in side the containers. Basic goal of the sensor node is the data collection and data gathering. So sensor node is tiny circuit that senses the environment and gathers the data. According to latest market research reports,the value of the sensor node market will reach to USD 38.41 billion and will grow at CAGR of 42.08 between 2016-2022. The important reason behind the growth of the IoT sensors, market is the manufacturing of low cost, intelligent, and tiny sensors, expanding the market for intelligent devices and wearables, requirement for real-time applications, government initiatives, deployment of IPv6 and RFID, and the role of sensor mixing idea for IoT sensors market.

According to markets, major companies involved in the development of IoT sensors include Texas Instruments Incorporated (U.S.), STMicroelectronics N.V. (Switzerland), TE Connectivity, Inc. (Switzerland), NXP Semiconductors N.V. (Netherlands), Broad com Limited (U.S.), Robert Bosch GmbH (Germany), InvenSense, Inc. (U.S.), Infineon Technologies (Germany), ARM Holdings Plc. (U.K.), Omron Corporation (Japan), Sensirion AG (Switzerland), Analog Devices, Inc. (U.S.), SmartThings, Inc. (U.S.), and KONUX Inc. (Germany) among others.

Sensor network can be also of two types:

1. Wired sensor network

2. Wireless sensor network

Selection between wireless and wired depends on convenience, interface, adoption, capacity and power availability. Famous protocol standards available for the sensor networks are RFID , ZigBee, Bluetooth, WI-FI, Cellular Networks, Z-Wave and Low Energy Bluetooth. The major challenge in the sensor network research is to take care about security, privacy, connectivity, power management, node failure, mobility and intelligence.

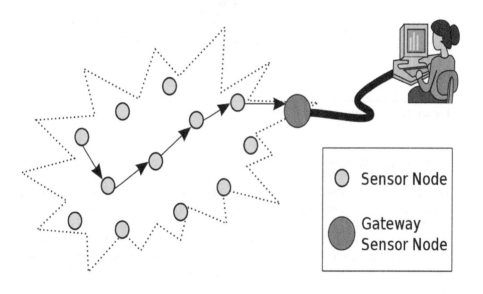

Figure 1.6: Simple Sensor Node scenario.

Figure 1.6 shows basic scenarios of sensor connectivity. Major applications for sensor networks include location monitoring, land slide detection, pollution detection, fire detection, water quality and water level monitoring, data center monitoring and data logging, health care monitoring, retail monitoring, food quality monitoring, and object tracking. Ubuntu Tinyos, LiteOS and ContikiOS[Chandra et al. (2016)] are the famous operating systems used for smooth sensor connectivity.

Figure 1.7 shows famous sensors used in various applications, and we can identify that sensors can provide easy and fast data collection services for IoT.

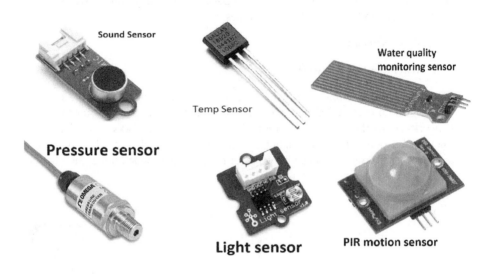

Figure 1.7: IoT sensors.

■ **Fog as a Service**:

Fog computing [Bonomi et al. (2012)] is the basic idea of CISCO (a famous networking company). Let your cloud computing functionality come closer to your devices [Yi, Qin, and Li (Yi et al.)]. Any devices that have a computing power, a storage facility and capability of connecting with the internet can be fog devices. Examples are industrial controllers, switches, router, video cameras, embedded devices. IDC estimates that more than 40% of data generated in IoT will be processed using fog device. Fog applications are part of IoT. They can gather the data, process the data and can provide human to machine interaction as well as machine to machine instruction. Fog computing is most useful in situations where data generated must be processed within very short time and high data is generated. Three types of services are expected from the fog as a service.

1. Low latency, less queuing delay(high storage capacity), low processing delay(high computing power), low transmission delay(high bit rate), better node

mobility and failure understanding[Summary, Latency, Rate, and Latencies (Summary et al.)]

2. High security, better node and device authentication, secure encryption and hash algorithms, secure cyber rules.

3. Efficient resource utilization, consume less power of the nodes

4. Handling heterogeneity, capable of handling heterogeneous devices.

A fog device must be equipped with following components:

■ Decision management

■ Authentication and authorization

■ Mobility management

■ System management

■ Resource management

■ Service management

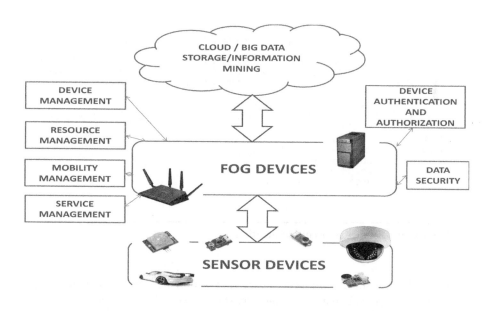

Figure 1.8: Fog as a service.

Based on Figure 1.8, we can identify that basic difference between cloud and fog devices is as follows:

■ Fog device receives data from sensor nodes, mobiles and other ground level nodes while cloud will receive data from various fog nodes.

■ Fog device provides response withing milliseconds while cloud may take 1-2 minutes.

■ Cloud stores the data for years while fog stores the data for transient periods.

So fog devices need to consider velocity, volume and variety of data and fog computing can be seen as a game changer service for IoT complexity.

■ **Communication and Data as Services**:

Major services provided by better communication facilities in IoT are reliable, fast and secure communication.

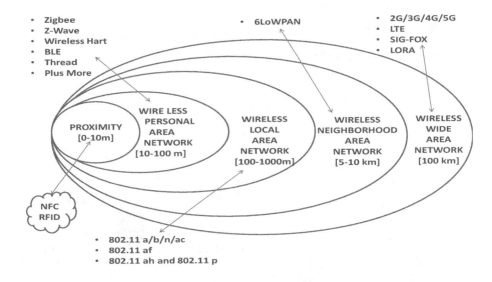

Figure 1.9: Distance wise communication.

Figure 1.9 was sourced from BCC research; it shows the major communication protocols used for communication from very short range to long ranges.

As shown in Figure 1.10, level wise communication can be understood as follows:

1. Level 1: Communication within sensor devices

2. Level 2: Communication between sensor devices and fog devices

3. Level 3: Communication between fog device and cloud

Figure 1.10: Level wise communication.

Another level of communication can be defined as a communication between user and cloud. Major challenges in communication in IoT compare to normal internet are as follows:

1. Communication within sensor devices is available, ex, coffee maker will communicate with bath shower, and bath shower will communicate with car, all are in same network.

2. Less human interaction in the communication, ex., car will inform office AC to start when it reaches near by some range.

3. Large number(growing dynamic) of inter-connected devices

4. More device triggered communication than human triggered communication

5. Different services than usual internet, human centered communication services to the device centered communication services

6. Dynamic traffic patterns and dynamic communication typologies

7. More loopholes for the cyber hackers

8. Route identification at each level

We can consider data as a service for the IoT in two types :

1. Normal data, mostly generated data by sensors and ground level sensors.

2. Control data, mostly a processed data either by fog device or by cloud data miners.

As discussed above, data in the IoT will be available in three major forms(data,information and knowledge)
Major challenges in the data of IoT are :

1. Heterogeneity of data

2. High volume of data

3. Heterogeneity of data

4. Security of data

5. Processing of data

6. Sharing of data

7. Access control of data

Well processed and secure data can provide expected signals for IoT devices.

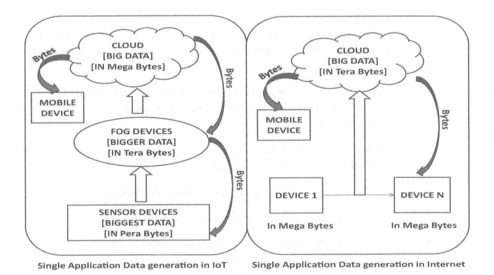

Single Application Data generation in IoT   Single Application Data generation in Internet

Figure 1.11: IoT Data generation.

As shown in Figure 1.11, data generation in the IoT will be totally different compared to traditional Internet.

■ **Cloud and mobile as a services:**

Major services provided by the cloud in the IoT will be as follows:

- Fog device identification, authentication and authorization

- Information collection from fog devices

- Information mining and knowledge generation

- Knowledge transmission to the mobile devices, mobile devices can be any users who are using application services

- Providing platform for the IoT application deployment, software for computing and development

Major challenges for the deployment of better cloud services in IoT are as follows [Sun et al. (2016)]:

- Heterogeneous devices

- Lack of standardization, starting from device to cloud

- Heterogeneous and homogeneous types of data

- Large number of devices and big size of data

- Highly expected accuracy

- Slow technology advancement

- Internet speed in many countries

- Active hacker groups

Development of mobile devices is the best invention of twentieth century. Rapid development in mobile device technology opens many doors for the IoT research community. Mobile devices in the IoT can be used for the collection of data as well as for the accessing of the data. Currently most of the smart phones have built in sensors like proximity sensors that are used to identify the distance between you and devices, accelerator/motion sensor used to identify motion of mobile phone, ambient light sensor to identify light availability, moisture sensors, gyroscope, compass, barometer to identify air pressure, WiFI protocols and so on. These capabilities of smart phones make them best platform that can be used for IoT deployment.

■ **Security as a service:**

Security in the IoT is the biggest challenge [Al-fuqaha et al. (2015)]. Rapid attacks on internet services have created a huge loss for many countries in terms of money as well as data. Major challenges that need to take care in IoT a security are as follows:

- Heterogeneity of sensor devices will lead to critical situations for the sensor device identification, inter-sensor device authentication and authorization

- Authentication between sensor nodes and fog devices

- Authentication between fog devices and cloud services

- Proper access control to the specified users

- Confidentiality of (may be use less) openly generated data

- Integrity of (may be useless) data

- Complexity of number of devices

- Less power availability with the sensor nodes

- Less computation capabilities with sensor nodes

- Maximum possibilities of node failure

Better implementation of security will increase the confidence of actual consumers of IoT services in the IoT.

1.3.5 Security Oriented

For security in the internet, there are three main pillars :

1. Confidentiality

2. Integrity

3. Availability

Later on, it was enhanced, and two more pillar were added: authenticity and non-repudiation. Confidentiality can be defined in very simple words as "Converting the data in such a form that except your receiver, nobody can read it". So confidentiality means applying a protocol to stop unauthorized access to particular data. Example: Leakage of personal financial information may lead towards life threatening situations. Integrity mean let us make sure that what ever data a sender want to send, the same data in the same size is received by the receiver. Integrity makes sure about the reliability of communication. Example: Change in the online prescription sent by doctor can lead to major reaction in patient. Availability means, let us make sure that legitimate user will have access of data for which he has permission. Authentication means, let us make sure that both the sender and receiver are certain about identity and availability of each other. While non-repudiation can be defined as an acceptance by sender that yes, he/she had sent the data, and they cannot deny it. Later on, they can't deny the authenticity of their communication [Anwar and Mahmood (2014),Lee and Lin (2014)].

Table 1.2: IoT generalized security requirements.

Requirement	Description
Confidentiality	Ensuring that only authorized users access the information.
Integrity	Ensuring no change in the actual data.
Availability	Ensuring that services will be available to authorized user on time.
Authentication	Both the sender and receiver are sure about identity of each other.
Auditability	Ability of system to continuously monitor the situation.
Trust-worthiness	Ensuring identity and trust of third party, if involved.
Non-repudiation	Ensuring occurrence and non-occurrence of action.
Privacy	Ensuring that everyone follows the predefined protocols and everyone maintains secrecy about their own work and data.
Accountability	Ability of system to hold users responsible for their action.

Table 1.2 discusses basic requirements for internet of things security. Continuous growth of IoT will result in 26 billion devices by 2020. These devices will be interconnected and will communicate with each other. So attackers will have larger ground in which to play. Report by IT major company, capgemini on "security in the IoT" says that password attacks will be highest threat on IoT products, and will be followed by identity spoofing, data modification, traffic sniffing and denial of service attacks. Seventy percent of industries believe that there will be a huge impact of security concerns on customers purchasing IoT products. Highest effect will be on manufacturing products followed by health and wearable products. Major cyber attacks have happened on wearable devices, followed by manufacturing, automation and wearable devices [Borgia (2014)].

Major reasons behind security concerns in the IoT are :

■ Limited resources to the nodes and devices

■ Temporary, homogeneous, large and heterogeneous network

■ Identity allocation, identity management and authentication of billions of devices

■ Authorization and access control

■ Accountability of policy implementation

We have derived security oriented vision as follows for the IoT :

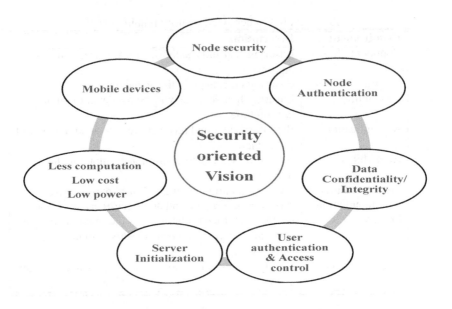

Figure 1.12: Security oriented vision.

As shown in Figure 1.12, we have defined security in four parts:

1. Device security

2. Data security

3. Service security

4. User security

1. **Device security:** Device security in the IoT can be for the any device involved in IoT deployment. It can be either RFID tag, RFID reader, sensor node, wearable device or any other computing device. Most of the devices in IoT suffer from physical attacks. In physical attacks devices will be directly damaged by attacker. Every time it is not necessary that attacker will be human only, it can be animal, environment or any other situation. Device also suffers from authentication attacks in which adversaries try to gain access of key, access of data transmitted or access of device itself by performing some attacks. Developing secure authentication algorithm is also one of the major challenges for the IoT. Authentication in the IoT can be shown by the following scenarios :

   ■ Authentication between sensor devices themselves

   ■ Authentication between sensor device and fog device

   ■ Authentication between sensor device and cloud

■ Authentication between cloud and user, user and gateway, user and sensor node

Devices in the IoT are low powered, so they suffer from battery power based attacks that include denial of service attacks or flooding attacks. Another major plus point for the devices is lower computation requirement, so sometimes when an attacker performs unrequired computation on the device, it will lead to a delay in output as well as battery consumption.

2. **Data security:** Basically data in the IoT may suffer from confidentiality break or integrity break. Passive attackers continuously observe the data transmission and do the analysis to gain useful information. Active attacker try to perform man-in-the-middle attacks and sometimes make changes in the data without any reason. Minor changes in the IoT problems can create big problems for the system. Example: 1 bit change can convert false to true or true to false in the input [Chahid et al. (2017)].

3. **Service security:** Services in the IoT can be attacked by the attacker in two ways:

■ Partially denied service
■ Fully denied service

Partially denied service attacks performed by attacker may not break complete system, but may modify the service. Example: modification of oxygen supply level in hospital service can become life threatening. Fully denied service means attackers stop the service by performing their attacks.

4. **User security:** There will be millions of users involved in the IoT. Attacks on the user application can be on authentication side, so attacker will try to attack during authentication or try to prevent authentication or try to perform MIIM attack. User attacks can be physical attacks when the user is the patient in the hospital or a logical attack when he is using generated knowledge for input [Khemissa and Tandjaoui (Khemissa and Tandjaoui)].

Edge nodes are devices that can be considered as RFID tags and computing nodes. Major attacks possible in IoT things are as follows:

Table 1.3: IoT Attacks : Edge computing

Location	Attack	Attack description
Edge computing node	Physical Attacks	Leads to total destruction or partial destruction of tag
	Denial of service	Battery draining using fake resource utilization, attacking on interface to stop functionality
	Hardware Trojan	Malicious modification of a device circuit which enables attackers to use circuit or access its functionalities to obtain access to data or software running on it
	Non-network side channel attack	Each device may release critical information during transmission, electro magnetic signal leaks from medical device can provide information about patient
	Camouflage	Hiding the edge level by inserting counterfeit or authorized node
	Corrupted Node/malicious Node	To gain unauthorized access of network on which device is working
	Battery draining	Physical or logical attack on battery by transmitting unrequired traffic
	Node replication attack	Attacker injects malicious node inside network by using identity of valid device

Table 1.4: IoT Attacks : RFID.

Location	Attack	Attack description
RFID Tag	Tracking Inventorying	Identify the RFID and track the object Examining RFID tag like EPC tag can give information about manufacturer code and product code
	Eavesdropping	To intercept, read and save the messages for future analysis and future attacks
	Physical attack and tempering	Physical attack on tag
	Relay attack	In a relay attack, an adversary acts as a man-in-the middle. An adversarial device is surreptitiously placed between a legitimate RFID tag and a reader. This device is able to intercept and modify the radio signal between the legitimate tag and reader.
	Counter feiting	Modifying identity of item
	Tag cloning	Replicating RFID tag by using its identity
	Denial of service	Jamming radio frequency channel so that reader can not read the tag
	Unauthorized tag reading	Adversary tries to read the tag and tries to fetch useful information
	Traffic Analysis	Analyzing traffic and use for the future attacks
	Side channel attack	Use pattern analysis for extracting information
	KILL Command	Permanently silence an RFID tag by partially erasing the tag data

Edge computing nodes: RFID readers, senors nodes, compact controlling nodes [Mohsen Nia and Jha (2016)]

Table 1.3 , Table 1.4[Mohsen Nia and Jha (2016)] discuss various attacks that are possible on gateways and RFID. Table 1.5 [Mohsen Nia and Jha (2016)] discuss possible attacks in the communication phase of IoT applications.

Table 1.5: IoT Attacks : Communication.

Location	Attack	Attack description
Communication	Eavesdropping	Also called as sniffing, Reading secret data exchange on communication link
	Side channel Attack	Reading the information and storing the information, which is unintentionally leaked
	Denial of Service attack	Jamming the transmission of radio signals
	Packet injection	Inserting or replicating node by malicious packets
	Routing - Black hole attack	Use malicious node to attracts all the traffic of network
	Routing - Gray hole attack	Advancement of black hole attack in which nodes selectively drop some packets
	Unauthorized conversation	Nodes will do communication with each other even though if they are not allowed

Table 1.6: Parameter and Countermeasure notations.

Parameter	Notation	Solution	Notation
Confidentiality ⇒	1	Side channel signal analysis ⇒	1
Integrity ⇒	2	Activate security Trojan ⇒	2
Availability ⇒	3	Intrusion detection system ⇒	3
Accountability ⇒	4	Securing firmware update ⇒	4
Audibility ⇒	5	Circuit/Design modification ⇒	5
Trustworthiness ⇒	6	Kill/sleep command ⇒	6
Non repudiation ⇒	7	Isolation ⇒	7
Privacy ⇒	8	Blocking ⇒	8
		Anonymous tag ⇒	9
		Distance estimation ⇒	10
		Personal Firewall ⇒	11
		Cryptography scheme ⇒	12
		Reliable routing ⇒	13
		De-patterning and De centralization ⇒	14
		Role based authorization ⇒	15
		Information flooding ⇒	16
		Pre testing ⇒	17
		Outlier detection ⇒	18

Table 1.6 [Sicari et al. (2015)] give prototypes of various parameters and security countermeasures, which are highlighted in Table 1.7[Mohsen Nia and Jha (2016)][Sicari et al. (2015)]

Table 1.7: IoT Attacks: Parameters and counter measures.

Attack	Affecting parameter	Counter measures
Dos attack	2,4,5,7,8	3,4,11,12
Side channel attack	1,5,7,8	5,6,7,8
Physical Attack	1-8	5
Node replication attack	1-8	12
Camouflage	1-8	4,12
Corrupted node	1-8	1,3,4,12
inventorying	7,8	6,7,8,9
tag cloning	1-8	6,7,8,10,12
counterfeiting	1-8	6,11,12
eavesdropping	1,7,8	6,7,8,11,12
Packet injection	2,5,6,7,8	3,12
Routing attack	1,2,4,7,8	14
Unauthorized conversation	1-8	3,15
Malicious injection	1-8	17
Integrity attack	1,2	18
tracking	7,8	6,7,8,9,14,16

Trojan attacks can be of two types, internally activated or externally activated. Externally activated trojan starts activity when it gets input from USB, sensor, antenna or any other external conditions, while internally activated attacks start its functionality when device meets certain conditions.

1.4   IoT Reference Architecture

1.4.1   Four Layered Architecture

After establishment of IoT idea, first layered architecture was proposed in [Mohsen Nia and Jha (2016)], consisting of three layers. First layer was wireless sensor layer, second layer was cloud servers, and third layer was application layer. Later on, five-layered architecture was proposed; first layer was edge nodes, second layer was object abstraction, third layer was service management, fourth layer was service compositions, and fifth layer was application layer. Microsoft also proposed reference architecture for IoT as shown in [Microsoft (2017)]. In 2014, CISCO had proposed seven-layered architecture. First layer in the CISCO model was edge nodes, second layer was communication, third layer was edge computing, fourth layer was data accumulation, fifth layer was data abstraction, sixth layer was application, and seventh layer was users and centers. Microsoft also discussed its view about IoT solution architecture.

Here, we will discuss four-layered IoT architecture, which is shown in Figure 1.13

■ **Layer 1: Physical Layer / Perception Layer / Object Layer**

■ This layer, we can call a "Things Layer". It consists of various things involved in the IoT. Commonly there are three types of things involved in IoT. First is individual sensors, second one is devices equipped with multiple sensors, and the third thing is controllers.

■ Basic functionality of this layer is to sense the environment, generate the data, and send the data to controllers as well as gateways

■ Individually deployed hundreds/thousands same or different sensors on campus, city, university, office, plant or any place that generates the first type of things for this layer, such as individual devices equipped with RFID tags used for authentication, access control, tracking, quantity check, identity, and supply chain management.

■ Example of second type of things is mobile devices, Such as fitness tracking devices. Devices that are equipped with multiple sensors. Mobile devices that have built in sensors like accelerometer, light sensors and many more. Wearable tracking devices have built in sensors that can measure heart rate, blood glucose, body temperature, blood pressure. This type also includes smart tattoos pasted on body to track various parameters. Environment and chemical sensors commonly equipped in smart hardware which can map air pollution, temperature, humidity, fog level, air pressure.

■ This layer provides "context awareness" to the devices. Context awareness means "activity based change awareness". There are various possibility when it comes to some decision based conclusions. We need input from many different sensors. Examples: we want to find weather the patient has a high blood pressure or not, then we need input from pressure sensor, temperature sensor, heart bit-rate sensor. So context awareness is very important.

■ Actuators are the device that convert the electric signal to action. Examples of actuators are lights, metros, speakers and controllers.

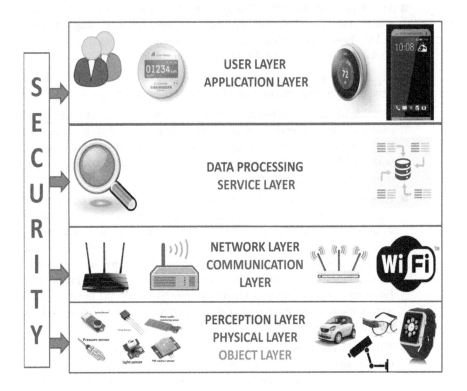

Figure 1.13: IoT layered architecture.

■ **Layer 2: Network Layer / Communication Layer**

- This layer combines mainly two functionalities: First functionality is secure transmission of data in complete IoT environment and second functionality is pre-processing using the concept called "fog computing (Word given by CISCO)" some time also called as "edge computing".

- This layer is a combination of various communication protocols used for all types of communication in the IoT environment. Examples of these protocols are given in Table 1.1. Short range protocols are used for communication between devices or with gateway, while long communication happens to communicate with cloud server or other network.

- We can say its motive is "Connecting things internally and externally".

- Fog computing performs pre-processing of data and converting data into information. It also provides temporary storage allocation and, performs security operations like encryption/decryption, authentication, device access control, data integrity and privacy. So we can also provide smart data pre-processing capabilities to pumps, motors, lights and so on.

■ This layer transfer information from perception layer to data processing layer.

■ Layer 3: Data processing Layer / Service Layer

■ This layer combines functionality of data base management, data mining, cloud computing, pattern recognition, recommender system.

■ This layer receives huge amount of data ("information") from network layer / communication layer; it does the processing on this information and generates the very useful "knowledge".

■ Services management and service composition for the service-oriented based architecture can be performed in this layer. Directory inside cloud should be updated and equipped with latest information so that when ever service is required, then IoT receives latest service.

■ Provides generation of real time and dynamic application from the available distributed applications.

■ This layer provides "context aware" computing; IoT follows completely dynamic environment.

■ This layer converts event-wise data into query based data; it generates database and schemes for filtering.

■ Layer 4: User Layer / Application Layer

■ This layer defines mainly two functionalities; the first is services to the direct user and second is applications deployed by using IoT environment.

■ "Use of traditional internet to use the service of IoT".

■ Reporting, analytic and designing of processed data to get knowledge from it for decision making. Decision input can be automated by device based on predefined threshold value.

■ Deployment and management of various applications based on domains, Examples: smart home, smart grid, smart agriculture, smart city, smart health care, smart manufacturing, smart transportation and smart logistics.

1.4.2   Seven layered architecture by CISCO

We will also discuss the seven layered architecture proposed by Cisco [You and Learn (2014)],

Cisco has proposed seven-layer architecture as shown in Figure 1.14:

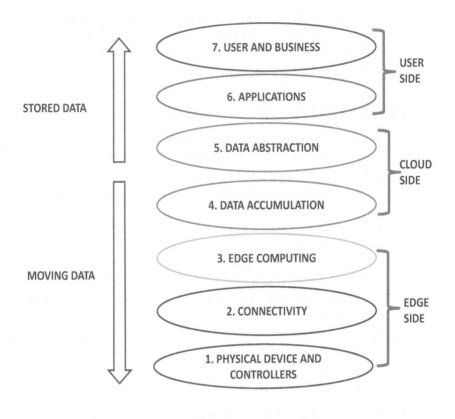

Figure 1.14: IoT layered architecture by Cisco.

■ **Layer 1: Physical Devices and Controllers**

  ■ This layer basically discusses about "things" in the IoT. Things can be devices or controllers. Devices basically collect the data and send it to controllers. Controllers forward the data to the valid destination at edge level.

  ■ Devices are also called computing nodes, Example: sensors, micro-controllers, micro processors, cameras, RFID Tags, RFID readers, blue-tooth devices, cars, wearable devices, intelligent machines and so on

  ■ This layer follows the short range protocols discussed in Table 1.1.

  ■ Security concerns in this layer are confidentiality and integrity of data, physical device security, authentication between devices, accountability and non repudiation of data transmitting devices.

■ **Layer 2 : Connectivity**

■ Communication and connectivity in the IoT is combination of currently existing standards as well as new standards. Due to heterogeneity of devices and non-availability of common standards, there is a big dilemma for connectivity.

■ Communication in the IoT can be :
   1. Communication between computing devices
   2. Communication between computing devices and controllers
   3. Communication between controllers and gateway as well as computing devices and gateway
   4. Communication between gateway and cloud
   5. Communication between users/application and cloud

■ This layer will use a combination of short-range and long-range protocols mentioned in 1.1 as the requirements.

■ Basic security concerns in this layers are all the possible attacks that happen in current internet scenarios. Some additional attacks like information leakage attacks are also possible.

■ **Layer 3 : Edge Computing**

■ Edge computing is also considered fog computing

■ Basic motive of this layer is to process the data at edge layer in such a way that it will reduce the computation overhead of the cloud.

■ It helps in traffic reduction and also accuracy in decision making.

■ This layer also increases possibilities of data loss and data theft.

■ No communication protocols are involved but, yes, some security parameters like physical security, dos and MitM need to take care.

■ **Layer 4 : Data Accumulation**

■ Storing the data, which is not going to be input for instant processing.

■ Whenever above layer needs some data for processing, then it can collect from this layer.

■ This layer will convert the data from packets to tables and schemes.

■ Conversion from event-based data generation to query-based data consumption.

■ Reducing the data through filtering and selective storing.

■ Event filtering, event comparison, event aggregation, event based rule evaluation.

◼ **Layer 5 : Data Abstraction**

- ◼ Rendering data and its storage in such a way that enable the development of simpler, performance-enhanced applications.
- ◼ Reconciling multiple data formats from different sources.
- ◼ Complete data to high level application.
- ◼ Securing data with appropriate authentication and access control.
- ◼ Normalization and denormalization of indexed data so it can help in faster access.
- ◼ So this layer basically converts the data in way that the application will need.

◼ **Layer 6 : Application**

- ◼ Reporting application, analytic application and controlling applications.
- ◼ This layer will communicate with data accumulation and abstraction layer.
- ◼ It is a very significant level for market and industrial requirements.

◼ **Layer 7 : Users and Businesses**

- ◼ This layer deals with the users and business owners that use the basic functionality; it will communicate with application layer.

1.4.3 IoT Nuts and Bolts

Nuts and bolts of any system define each and every small physical things involved for the successful implementation of any system. For the IoT, the major nuts and bolts are as follows:

1. Sensors
2. Wires (all the cable starting from small male-male jumper cables to fiber optics cable)
3. Wireless routers, switches
4. Small and large storage devices
5. Computing devices
6. Cameras,mobiles, led lights, signalling systems, breadboards

7. Micro-controllers, micro-processors

8. RFID tags, RFID readers, QR Codes, BLE Devices.

9. Physical things

10. Humans

So IoT will make the whole world a joint family and will prove to be a successful revolution after the internet.

1.5 IoT Security

Internet has proved itself a great revolution in the world of technology. After establishment of internet, the revolution took place. The World Wide Web and other entities came in to the picture, hundreds of communication protocols were developed and thousands of devices were manufactured. People started to use social networking sites for fun, banking sites for their financial transactions. But....behind all this, with the fun and facility, one other face of internet also came in the picture and that was the face of cyber hackers and attackers. Attackers or hackers can be either passive or active. The basic motive behind attacker's, and hacker's, expansion is either to do financial damage, social damage or defense damage. And from their internet facility and fun came the threats. Various ethical hackers, government organizations, private organization, researchers have developed thousands of ways to protect internet but still every day we see in news that hackers are damaging billions of financial transactions of particular persons, companies or countries. In the world of internet, the journey from 2 devices to 20 billion devices was not easy due to these attacks and thefts. There are many reasons behind attacks in the internet(not in IOT), first reason is not availability of fully secured algorithms, second reason is lack of knowledge about security rules by people and third reason is intelligence of hackers and attackers.

1.5.1 An overview

As of now, various communication protocols like HTTP + SSL, TLS are secured through various algorithms like AES, DES, MD5, HMAC, SHA and so on. But the major question in the IoT is

"Whether all of these algorithms applied in traditional internet are sufficient to provide security in the IoT? If YES, then try it and if NO then why?"

After establishment of internet connected photo sensors to a soft drink vending machine at Carnegie Mellon University by students, in 1980, we can say the journey of IoT started. CISCO predicted that 50 billion devices would be connected with each other via internet in 2020. Rapid expansion of devices makes life easy for people as well as attackers and hackers. Machine research published one report in 2015; it noted that total numbers of interconnected machines will be 5 billion in 2014 to 27 billion in 2024. Other organizations like Gartner believe 6.4 billion devices would be in use, IDC says 9 billion devices and Juniper researchers estimate 16 billion devices by 2022. So again, let us come back to our original question that will the security aspect, communication aspects and connectivity aspects of internet and IoT be the same?

As we believe, a big NO, it will never be the same. Various parameters that make significant differences between internet and IoT.

1. Huge number of resource constrained devices.

2. High storage space requirement at device level as well as cloud

3. Huge dynamicity in topology

4. High security constraints

Most of the devices, we are going to use in the IoT are not the same devices that we are using in the internet. In the internet, data generator was human while in the IoT, the data generator will be human as well as the devices. In the internet, devices did not have decision power or cognitive capabilities, but in the IoT, devices are going to have decision power as well as cognitive capabilities, learning, smelling and watching. Heterogeneity of the devices creates heterogeneity of communication protocols and communication hardware; heterogeneity will create big security risk for people.

Basically in security, we have gone through a triangle of parameters called CIA (confidentiality, integrity and availability). Confidentiality makes sure that no third party was able to read or understand actual data. In the cryptography, confidentiality was achieved by various encryption algorithms like AES (Advanced Encryption Standards) and DES (Data Encryption Standards). Integrity can be defined as a prevention of modification of content of data as well as size of data by any third entity. Integrity in the internet was achieved by hash functions like SHA (Secure Hash Algorithms) and MD5 (Message Digest 5). Availability means whenever an authentic user wants a particular service, it should be available for him to use. Availability can be achieved by various access control mechanisms like user name and passwords and predefined protocol enforcement. But for the IoT, various researchers have promoted more parameters like authentication, availability, accountability, non-repudiation, privacy, trust, reliability, scalability, device anonymity, data anonymity and so on [Sicari et al. (2015)].

Table 1.8: Traditional Internet vs. IoT.

Parameter	Internet	IoT
Environment	Close and secure	Open and hostile
Device Identity	IPv4 Address	RFID tag, BLE number,NFC number,IPv6 Address
Number of devices	Limited (Maximum 2^{32})	Not Limited(Approx. 50 billion by 2022)
Types of devices	Computers, Servers, Routers	Computers, Servers, Routers, Mobiles, RFID, Sensors, Micro controllers, Embedded Devices
Device Avg. RAM	1 GB to 16 GB	64 Bytes to 1 GB
Device Avg. ROM	512 GB to 2 TB	SD Card(16 to 128 GB)
Device Avg. clock rate	2.5 to 3.5 GHz	12 MHz to 1.3 GHz
communication	Maximum Human to Machine	Human to Machine,Machine to Machine, Human to Human
latency	In millisecond	In microsecond
throughput	80% Average	100% Expected
Application layer protocol	HTTP, DNS, DHCP, SNMP, etc.	HTTP, DNS, DHCP, SNMP, MQTT, COAP, XMPP, etc.
Transport layer protocol	TCP and UDP	TCP and UDP
Network layer protocol	IP,TCP	AODV,IP,6LoWPAN,DTLS
Physical/Mac layer protocol	Medium access control	NFC, Bluetooth, ZIgBee , Wireless-HART, Z-Wave
Cryptography	Traditional Algorithms	Light weight Algorithms
Power	Unlimited	Very Limited
Data generator	Humans	Devices
Data Volume	In terabytes	In Zetabytes

Above table discusses basic comparison between internet and internet of things so it can help us to understand upcoming challenges.

As already discussed, for the IoT, the following will be security targets :

■ Device security

■ Data security

■ Service security

■ User security

Device security belongs to first layer of IoT architecture 1.13. Data security will be in part of network layer and data processing later. Service security will be also part of network layer and data processing layer. User security need to take care at application layer or user layer. As discussed in Borgia (2014), IoT security also be said as:

- Secure data collection

- Secure data transmission

- Secure data storage

- Secure data access

If we compare wireless sensor network and IoT, then we can say wireless sensor network is the subset of IoT IPv6 and RFID are used in the IoT for the device identification, while in the WSN, the device can be identified by IPv4 based Wi-Fi. In the traditional wireless sensor network, we use TCP or UDP as a data generation stream protocol, while in IoT we use data-gram transport layer security protocol. In the traditional internet based WSN, we use HTTP as a application layer protocol, while in the IoT, we use constrained application protocol and message queuing telemetry transport(MQTT).

"Do you think in the IoT, only key sharing will be a problem for authentication?","You don't think that algorithm sharing may be also required?", A question in the UN-STANDARDIZED ENVIRONMENT OF IoT.

Traditional cryptography algorithms use resources (battery, bandwidth, storage) in the following tasks:

- Mathematical computations

- Input size

- Key size

- Number of rounds

- Data aggregation

- Output forwarding

Mobility management in the IoT will be difficult compared to internet due to lack of proper identity and communication standardization. Nowadays device identities in the IoT are not allocated based on particular fixed protocols. So it will become very difficult to identify host agent and the foreign agent for the particular device. And this will create security problems like authentication of the device, identification of the device and privacy.

Lightweight cryptography is discussed in section 1.7. Lightweight cryptography can lead towards IoT security from internet security. In this section we discuss security architecture, in which we have shown, the communication parameters that need to take care at each layer of IoT four-layer architecture. We have explored each parameter and cited some of the schemes proposed for each security parameter that must be focused in IoT security [Khemissa and Tandjaoui (Khemissa and Tandjaoui)].

1.5.2 An Architecture

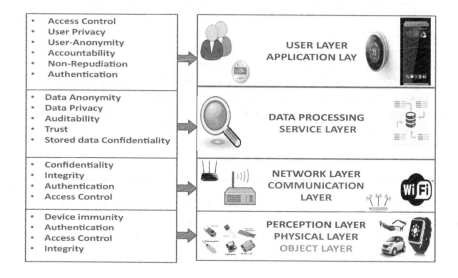

Figure 1.15: Layered Security Architecture.

Figure 1.15 shows basic layered architecture depicting which security parameters we will need to consider while designing security for the internet of things.

For the **perception layer**, we will need to consider for authentication, access control, device immunity and integrity. Authentication can be defined as, "Both: sender and receiver must be sure that "from and to" which user/entity they are communicating are the actual ones with whom he/she wants to communicate". In the IoT, we need to consider **authentication**, for the following two entities [Flauzac et al. (2015), Mundhenk et al. (2015)]:

1. Device authentication

2. User authentication

Authentication in the internet will be performed in the IoT between :

■ Device to device authentication

■ Device to user authentication

■ User to device authentication

■ Machine to Machine

A simple authentication model for the IoT is discussed in section 1.6. In the IoT, due to high mobility, devices need to authenticate user and device which do not belong to same network, the, first time they identify the network. A successful authentication makes sure that all three basic parameters for security (confidentiality, integrity and availability) are satisfied. There are various schemes developed for the authentication of RFID tags, other tags EPSGlobal, class-1, Generation-2 and ISO/IEC 18000-3 tag provide authentication of read and write operation by password. Public key infrastructure and key sharing using third party entity like registration server can create more complexity in authentication due to lack of resources. Other challenges in authentication at perception or physical layer are node failure due to power failure or physical attack, node failure due to environment effect, node authentication failed due to congestion of data in the channel, failure due to collision of control signal and data signal.

"Most of the time, when we calculate round trip time of the packet, we ignores queuing delay.",In the case of IoT, will it be possible or identical....Just think on it. "....tick.....tick...."Thousand of devices, limited gateway storage capacity......"

In the perception layer, the other major challenge is **access control**. Access control means, which user or device will now have how much access of the other device's and user's data. For the patient, family doctor will have access to complete health history of patient, but orthopedic doctor will not have access to complete history. At the perception layer, whenever a user from the application layer requests access to sensor node or camera, then how much information should be allowed that can be said as an access control. Currently most of the smart phones have App-Lock functionality; that functionality is called an access control.

Access control can be secured by password. There are many schemes developed for access control like role-based access control, attributed based access control, capability-based access control and so on. So access control is the permission assigned for accessing resources to different users and devices. Device immunity means devices must be able to save and recover itself from physical attacks in most cases.

Integrity in the perception layer means whatever data transmissions happen between devices or device and user must not be modified. Integrity at this layer becomes a critical point because if data is modified by attacker then all the subsequent layers will have modified data and will lead to false knowledge generation.

At network layer where fog devices and cloud storage are available, need to consider, confidentiality, integrity, authentication and access control. Confidentiality becomes a major issue here due to availability of internet and public domain. Data confidentiality could be discussed in the perception layer also, but due to availability of local network assumptions,we haven't discussed it. Authentication in this layer is required between following entities :

■ Sensor node and device gateway or fog device

■ Sensor node and user via gateway or fog device

■ User and cloud server via gateways

■ Machine to cloud and machine to gateway

Access control at this layer becomes a very important aspect due to big data available at this layer. Between perception layer and network layer, packets will travel from many different nodes. End-to-end authentication and access control need to be considered. Data about a person's health , data about the person's job, data about their food habits, and data about a person's shopping habits should be considered. All this data will be in one place and various entities like doctor of patient, businesses, and analyst will try to access the data. Which entities will have how much access of which data? That is called actual access control.

Many other access control methods like extended-role based access control, discretionary access control, mandatory access control, access control list and other methods based on physical unclonable functions have been developed. Designing access control mechanisms includes various parameters like delegation support, granularity, access right revocation, time efficiency, scalability and security. [Sfar et al. (2017)] have discussed role-based access control that ensures authentication and access control. Authors have discussed capability-based access control, that is defined it as a "token, ticket or key that gives rights to processor for accessing certain resources. [Sfar et al. (2017)] have discussed trust-based access control and fuzzy trust-based access control. In trust-based mechanism trust value depends on experience, knowledge and recommendations [Flauzac et al. (2015)].

Data processing layer OR service layer deals with processing of gathered data and providing input for management and composition of various services. At this layer, we are required to deal with data anonymity, data privacy, audibility, trust, stored data confidentiality. Trust creation between intelligent devices and technology, user and intelligent devices, user and technology become important aspects for the ecosystem of the IoT. Trust management can be handled based on past performance of particular node, past activities of node. There are various algorithms and schemes developed for the trust management. Major challenge for the trust mechanism comes when interaction happens between unfamiliar and surprised nodes. So we need separate protocols for the humans, devices and services in the trust management. Degree of trust defines dependency level.

Trust management is defined as a approach to making decisions about interacting with something or someone we do not know, establishing whether we should proceed with the interaction or not. Trust management can differentiate in two types. The first one is deterministic trust in which some policy or certification is defined for the trust establishment, while non-deterministic trust management depends on some dynamic parameters like recommendation, repudiation, prediction and social network. Policy based trust management will define a policy to find the minimum trust level of nodes.

Certificate-based trust management will lead towards existence of certificate authority, which will provide certificate of trust to communicating entities. Recommendations based on either explicit or transitive recommendations use prior information about the other communicating party. Prediction-based trust is very risky, and it comes into the picture when there is no prior information available about counterpart. Reputation based access control considers global repudiation of the parties.

Privacy for the IoT is, "It is the fundamental right of the user/person to decide how much information about him/her will be collected, when the information will be collected, where the information will be collected, how the information will be collected. At the same time the user also has the right to decide how much data will be shared with whom, why and how or will be used by whom, where, why and how". Because in the IoT, data collection, data mining and data provisioning will be completely different than the internet. W3C has developed one platform called a platform for privacy preferences to connect with it.

For example, the user visits a certain place and CCTV cameras are there. Now if this user doesn't want to allow his picture to be seen by everyone who has access to CCTV, but he wants only investigating agencies to see his clear picture, then everyone else can only see his blurred image. That is the privacy control in the IoT. Currently very few systems will have privacy rights. Various privacy preserving schemes are discussed and developed by researchers in [Sicari et al. (2015)], data tagging based privacy preservation technique is discussed. Authors have also proposed user-controlled privacy preserved access control protocols. In [Sicari et al. (2015)] discretionary access and limited access based privacy schemes are discussed. While in [Sicari et al. (2015)], Continuously anonymizing streaming data via adaptive clustering system is proposed. [Sicari et al. (2015)] has discussed key privacy preserving using attribute based encryption, Key policy attribute based encryption and ciphertext-policy attribute based encryption schemes are discussed. Author in [Sfar et al. (2017)] discussed privacy solutions based on data privacy and access privacy, data privacy can be anonymization based and encryption based privacy solution [Hassan et al. (2013)].

[Sfar et al. (2017)] have discussed three schemes: K-anonymity, L-diversity, and T-closeness. K-anonymity is used to improve location privacy. Authors haves also defined that L-diversity can overcome heterogeneity attack and background knowledge attacks in K-anonymity. [Sfar et al. (2017)] have discussed T-closeness for data privacy and its say that distribution of sensitive attributes in any group should be close to their distribution in the overall database. [Dusart and Traor (Dusart and Traor)] have discussed various light weight algorithms based on ANF and XOR operations. Hash function also deals with privacy due to its capabilities of easy computation, collision resistant, and pre-image resistant. [Leander, Paar, Poschmann, Schramm, and Horst (Leander et al.)] have discussed various elliptic curve based and hyper elliptic curve based cryptography schemes. Due to its algebraic nature, an elliptic curve can be very useful in resource constrained environments. "KILL" command in the RFID can be used for access privacy. Light weight protocols can be useful in privacy preserving and authentication. Data anonymity means systems need to consider who is the actual generator of the data and from where this data came through. This type of data processing is already done on the fog device, and so on.

User layer and application layer deal parameters called access control, user privacy, user anonymity, accountability, non-repudiation and authentication. User anonymity and device anonymity mechanisms need proper device and user identification system. Various parameters for the detection of users includes its bio metric, user name, password, smart card. Identity of the device can be RFID tag, ID and smart cards. Proper authentication mechanisms make sure that user anonymity can be removed. Only authenticated and registered users will be able to communicate.

So IoT security is far different than traditional internet security where confidentiality, integrity and availability matter but also authentication, access control, privacy, identification, reliability, trust, heterogeneity are also very important parameters that need to be secure to prepare foolproof(near to impossible, but not impossible) secure IoT architecture.

1.6   IoT Authentication Models

Authentication, in the world of cryptography authentication came into the picture when Lam port in 1981, published one time password based authentication technique. Authentication attracted many researchers for various reasons. A successful authentication makes sure that confidentiality, integrity and availability are secured. And this is the reason various authentication schemes came in to the picture. Here for the IoT, we can think of basically two authentication models. In the short range authentication model, device to device authentication using RFID or BLE can happen[Janbabaei et al. (2016)].

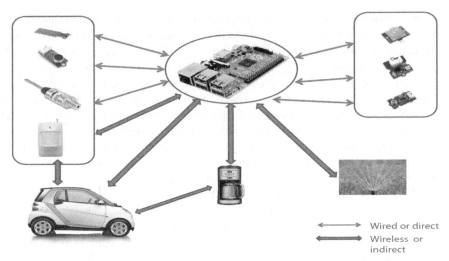

DEVICE TO DEVICE AUTHENTICATION [SHORT RANGE APPROACH]

Figure 1.16: Short range authentication model.

In Figure 1.16 we have shown a communication approach for short distances in the IoT. Here devices perform communication with other devices via micro-controller or micro-processors(also a device). So the authentication model between device to device will come into the picture. Most of the devices in the IoT are RFID tag enabled devices, so many researchers have prepared authentication schemes using RFID tag as an identity.

As discussed in Figure 1.17, long range authentication model represents complete end-to-end authentication scenario in the IoT. Here device to device authentication is already part of it, but it also adds user to device and user to user authentication also. And this is the actual heterogeneity of the internet where everyone will try to authenticate every other user. So it becomes important to think about whether normal authentication schemes with encryption algorithms like DES or AES will be suitable for resource constrained users and devices or will there be a need for any other solution like light weight authentication algorithms.

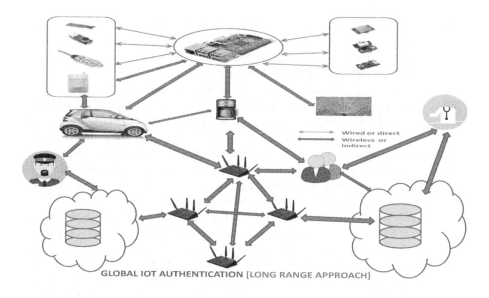

GLOBAL IOT AUTHENTICATION [LONG RANGE APPROACH]

Figure 1.17: Long range Authentication Model.

1.7 Light weight cryptography

1.7.1 An overview

A need for light weight authentication was generated due to various constraints of communicating parties as well as network came into the picture during deployment of IoT. Major constraints in IoT deployment created difficulty because of the types of billions of devices involved in communication. These devices are not the general networking devices that we use in traditional internet. Most of the devices involved in IoT are embedded and tiny devices. They have various resource constraints like they don't have long battery backup, they don't have high storage capacity, and they don't have that much high computing power. So these devices generated a need for developing light weight (in energy, storage and computation) cryptography algorithms. There is a vast difference between traditional wireless sensor networks and IoT. Most of the nodes in the IoT suffer from node impersonation attacks, and they will create huge loss of data [Li and Liu (2017)].

A report prepared by the National Institute of Standard and Technology in 2014 [McKay et al. (2017)], titled "Report on Light weight cryptography" have discussed many different approaches for light weight cryptography. Devices involved in conventional cryptography were servers and desktops while light weight cryptography devices are RFID Tag, embedded devices. Most of the micro-controllers like NXP

RS08, ARDUINO, COP912C have RAM of 64 bytes or less. Most of the RFID and sensors are not battery powered. So these types of devices must use lesser amount of gate equivalents(GE). Gate equivalent is the area that is required by two input NAND gates, so area in GE is measured by dividing the area in μm^2.

Cryptography algorithm performance can be measured based on power and energy, consumption, latency, and throughput. The resources required by particular hardware can be measured using gate area or GE. RFID tags are of two types, one which generates power from reader and the other is battery powered. But in both the conditional power consumption need to be considered. Power consumption also depends on other parameters like threshold voltage, clock frequency and technology used. Latency of response and throughput of process are also important matrices of performance computations. Comparing the number of GE used in different technologies can't be directly compared because it is very specific for individual technologies. A low cost RFID can have total gate count of 1000 to 10000, out of that only 200 to 2000 can be used for security. Software resources used by algorithms can be counted by the number of registers used, number of bytes of RAM and ROM required. So it is not difficult to secure the IoT, but actual difficulty lies in maintaining balance between security, performance and resource requirements [McKay et al. (2017)].

1.7.2 A literature study

Block Ciphers [Bogdanov, Knudsen, Leander, Paar, and Poschmann (Bogdanov et al.)]

Lightweight Data Encryption standard [Leander, Paar, Poschmann, Schramm, and Horst (Leander et al.)] To enable use of DES in the RFID technology, an advanced low weight version of DES was prepared. Only one s-box was used instead of 8 s-boxes in original DES. Initial permutations used in DES were completely excluded in DESL. Key size in DESL is just 56 bits and brute force algorithm took three days to break it, so it is suitable for short time protection. It uses 20% less gate equivalences and 25% less average current than DES.

AES Based lightweight authentication protocol [McKay et al. (2017)] Authors tried to introduce nonce based advanced light weight encryption standard algorithms. But except that 105 compared to 226 in AES, nothing was reduced. Gate equivalent taken by ALE was 2579 compared to 2435, while power consumption was 94.87 compared to 87.84W.

PRESENT PRESENT was the first light weight cipher developed for resource constraint devices. PRESENT has input plain text of 64 bit and key size was 80 and 128 bits. It took 32 cycle per block, 200 kbps throughput at 100 KHz. Gate equivalent was $1570\mu m^2$ [McKay et al. (2017)].

RC5 RC5 was the light version of RC4 and its previous versions. It allowed 32, 64, and 128 bit block size, 0 to 2040 key size and the number of rounds involved was 0 to 255 [Rivest (Rivest)].

Major parameters considered during the design of light-weight block ciphers were smaller block size, smaller key size, simple rounds(fewer s-boxes requires fewer GE). Simple key schedule(we can use secure key derivation function).

Hash Functions

SPONGENT SPONGENT was the hash function with the smallest footprint. It uses key size(bits) of 80, 120, 144, 208, 240, hash generated(bits) of 88, 128, 160, 224, 256, gate equivalent of 738, 1060, 1329, 1728, 1950, power consumption(μW) 1.57, 2.20, 2.85, 3.73 and 4.21, throughput(kbps) 0.81, 0.34, 0.40, 0.22, 0.17 respectively [Bogdanov, Knudsen, Leander, Paar, and Poschmann (Bogdanov et al.)].

LHASH: LHASH supports three different digest sizes: 80, 96 and 128 bits, provides preimage security from 64 to 120 bits, second preimage and collision security from 40 to 60 bits. LHash requires about 817, 1028 GE. In faster implementations based, LHash requires 989, 1200 GE with 54, 72 cycles per block. Power per bit consumed was 34008, 56669, 11337, throughput was 2.40, 1.81, and 0.91, and cycles of 666 and 882[Wu et al. (2013)].

PHOTON PHOTON hash function was developed for RFID tag. PHOTON generates hash size of 80, 128 bits,use GE of 108 and 996, Power per bit of 30621 and 69845, throughput 2.82 and 1.61, cycles of 708 and 996 [Guo et al. (2011)].

Quark: In 2010, author proposed an other lightweight hash function. It generates hash size of 128 bit, GE of 1379, power per bit 93772, throughput 1.47, and cycle of 544 [Aumasson et al. (2012)]. Secure hash functions must be secure against preimage, second pre-image, and collision attack. It must have smaller internal state and output size. In light weight hash function, input sizes are smaller (at most 256) bits compared to traditional hash functions.

Stream Cipher: e-Stream competition organized with the aim of identifying new stream ciphers by European network for excellence for cryptography. A famous light weight stream ciphers are **A5/1, ChaCha, E0, GRAIN, MICKEY V2, SNOW 3G, Trivium:**. In the many research documents authors have given comparative analysis of various light weight stream ciphers. ISO/IEC 29192, light weight cryptography is a famous standard for light weight cryptography. [Amin et al. (2016)] discussed a light weight authentication protocol using smart card. [He and Zeadally (2015)] have given analysis of various authentication schemes developed using elliptic curve cryptography in the health care environment. Another light weight authentication schemes is proposed using nonce and EX-OR operation, by authors in [Khemissa and Tandjaoui (Khemissa and Tandjaoui)].

Most of the light weight authentication schemes are developed using RFID as a identity and EX-OR and multiplication based mathematics like elliptic curve cryptography. We haven't discussed all the schemes due to limited scope of this book.

1.8   IoT historical approach

Communication started its journey when our ancestors used pigeons for the long distance communication. Later on in 1830, land line communication started its journey. In 1957, the first satellite Sputnik-I was ready to fly and the first radio wave transmission took place. In 1962, defense advance research project agency started the journey of the internet by connecting four different nodes. In 1969, advance research project agency used TCP/IP model for the first time for communication. In 1983 IETF represented a famous communication protocol IPv4, and in 1995, IPv6 came into the picture. 1993 was the year when global positioning system was established. In 1999, Kevin Ashton, an executive director of Auto-id lab of MIT used the word IoT. Later on many entities and researchers started to think about connecting things. Cloud computing, big data, machine learning, recommender systems, neural networks have added some tasty ingredients in the development of IoT. As discussed in the chapter, many researchers and organizations are involved in the deployment and development of the IoT.

Major constraints in the IoT concerns global standardization. There are many organizations involved in the development of global standards for the IoT, but globally accepted standards are still not decided. OCF (Open Connectivity Foundations), ETSI (European Telecommunication Standards Institute), IEEE, IETF(Internet Engineering Task Force), IOTA, IBM, Microsoft, CISCO, TIA(Telecommunication industry association), ATIS(Alliance for Telecommunication Industry Solutions) or oneM2M have tried to prepare standards for IoT either individually or collaboratively. IoTivity is the open source project funded by open connectivity foundation for the development of IoT stanrds. ETSI has two entities: one ETSI M2M basically focuses on development of machine to machine services, functional requirements, interfaces and architecture. The other entity ETSI ITS (Intelligent transport system) focus on the development of standardization of vehicle to vehicle communication.

An internet engineering task force works for the development of communication protocols for internet-based technologies. Famous IoT protocols like COAP, CORE and 6LoWPAN are developed by IETF. IEEE community also contributed to the development of various IoT related standards like IEEE 802.11, IEEE 16092/3, IEEE 802.16p and so on. IoT-A proposed architectural reference models for IoT.

Around 1900, in Egypt, the first symbols of cryptography were found where the aim was to dignify the message. Many people still believe that a founder of Indian "Arthshashtra", chanakya was aware of cryptography, and he had used the cryptography but still there is not strong evidence available for that. In 100 BC, Julius Caesar used the substitution method to develop first cryptography cipher. In the 16th century, Vigenere developed the first key based cipher where the key was added in plain text. In 1900 the Hebern Rotor Machine was founded where key was embedded in a rotating disc. The engima machine, which used more than one rotor, was during World War II by the German army. Up to the second world war cryptography was used for military purposes only. In 1973, NIST proposed development of more complicated design in the world of cryptography and developed first block cipher Lucipher, later called a data encryption standard. In 2000, NIST published an advanced encryption standard, a famous and basic principle of cryptography known as Kerckhoffs's principle, "The secrecy of your message should always depend on the secrecy of the key, and not on the secrecy of the encryption system".

1.9 IoT futuristic approach

As per global reports by scientific communities, we will have 212 billion devices by 2020, 45% internet traffic by machines in 2022, and a $6. 2 billion economic impact by 2025 from IoT. So this growth of devices will create immense opportunity for the various businesses to grow exponentially. Adoption of the growing technology will create revenue opportunity. Manufacturing industry will grow with smart manufacturing, health industry will grow with wearable patient tracking devices. Agriculture will grow with adoption of smart agriculture in which smart farming and smart irrigation system increase crop production. Intelligent transport systems will create high impact on transportation and logistics . Retail industries will grow with smart product tracking and smart billing systems. This technology will enhance people's lives in terms of comfort and smooth life. But at the same time these changes will create huge opportunistic challenges for the technology developer. Expansion of the number of devices will create huge heterogeneity, complex topology, high congestion, long collision, and many security loopholes. Major futuristic developments in the IoT will be required for following fields :

- Standardization and common architecture for hardware,software and protocols

- Quick and reliable communication

- Battery, computing and identity management for devices

- Data handling using secure data mining, context aware computing

- Security

Web 2.0 application has added colors in the internet. Similarly web squared will also add colors in the IoT. It enables integration of web and sensing technology, so integration of web squared and web 2.0 will change the future world. Social IoT is the integration of IoT and Social Network.

- Social group of objects from same manufacturer

- Social group of objects from same place

- Social group of objects from same application

- Social group of objects from same user

- Social group of objects meet by chance

Context aware computing is also a big challenge that we need to work on.

IoT security will need low power and low area aspects of IoT. Lack of proper cyber policies and its enforcement, uncertain capability of devices, lack of global suitable light wight protocols will be major challenges for IoT security. Biometric-based system will create opportunity for context aware computing and high secure systems. Industrial IoT version 4.0 will need various secure cyber policies and protocol.

1.10 Summary

In this chapter we have discussed ground level basics for the IoT, we have started with an overview of IoT. We have basic IoT architecture. We have discussed application, thing, communication and research-oriented vision. The basic idea was to discuss the vision to showcase in which direction the community is working. In research oriented vision we have discussed sensor, fog, communication, data, cloud, mobile and security as service concept. We have also discussed various attacks that can be caused in RFID, edge computing and communication. While we develop any new scheme, we should be aware of any attacks. In the following section, we discussed two types of architecture. One was a four layered architecture and other a seven layered architecture by CISCO. In next section, we discussed security architecture for the same four layered architecture. We have also discussed internet vs. IoT comparison table. So the readers can have clear differentiation for that. For the authentication model, we discussed short-range and long-range authentication models, followed by historical and futuristic approaches of the IoT. We have tried to teach ground level basics for the IoT so any first-time reader can easily understand about IoT, the importance of security in IoT, and the importance of authentication in IoT.

1.11 References

Abouzahir, S., M. Sadik, and E. Sabir (2017). Iot-empowered smart agriculture: A real-time light-weight embedded segmentation system. In E. Sabir, A. García Armada, M. Ghogho, and M. Debbah (Eds.), *Ubiquitous Networking*, Cham, pp. 319–332. Springer International Publishing.

Al-fuqaha, A., S. Member, M. Guizani, M. Mohammadi, and S. Member (2015). Internet of Things : A Survey on Enabling. *17*(4), 2347–2376.

Aloul, F., A. R. Al-ali, R. Al-dalky, and M. Al-mardini (2012). Smart Grid Security : Threats , Vulnerabilities and Solutions. (971).

Amin, R., N. Kumar, G. P. Biswas, R. Iqbal, and V. Chang (2016). A light weight authentication protocol for IoT-enabled devices in distributed Cloud Computing environment. *Future Generation Computer Systems*.

Anwar, A. and A. N. Mahmood (2014). Cyber Security of Smart Grid Infrastructure Abstract :. (January).

Ashton, K. (2009). That 'Internet of Things' Thing. *RFiD Journal*, 4986.

Atzori, L., A. Iera, and G. Morabito (2010). The Internet of Things: A survey. *Computer Networks 54*(15), 2787–2805.

Aumasson, J.-p., L. Henzen, and W. Meier (2012). Quark : a lightweight hash âĹŮ. pp. 1–24.

Bogdanov, A., L. R. Knudsen, G. Leander, C. Paar, and A. Poschmann. PRESENT : An Ultra-Lightweight Block Cipher.

Bonomi, F., R. Milito, J. Zhu, and S. Addepalli (2012). Fog Computing and Its Role in the Internet of Things Characterization of Fog Computing. pp. 13–15.

Borgia, E. (2014). The internet of things vision: Key features, applications and open issues. *Computer Communications 54*, 1–31.

C, Muthu Ramya, Shanmugaraj.M, P. (2011). STUDY ON ZIGBEE TECHNOL-OGY. *IEEE*, 297–301.

Chahid, Y., M. Benabdellah, and A. Azizi (2017). Internet of Things Security. *IEEE*.

Chandra, T. B., P. Verma, and A. K. Dwivedi (2016). Operating systems for internet of things: A comparative study. In *Proceedings of the Second International Conference on Information and Communication Technology for Competitive Strategies*, ICTCS '16, New York, NY, USA, pp. 47:1–47:6. ACM.

Chaturvedi, M. and S. Srivastava (2017). Multi-modal design of an intelligent transportation system. *IEEE Trans. Intelligent Transportation Systems 18*(8), 2017–2027.

Chen, D., M. Nixon, and A. K. Mok (2014). WirelessHART and IEEE 802.15.4e. pp. 760–765.

Communication, N. F. (2011). Smartphones. pp. 4–7.

Dusart, P. and S. Traor. for Low-Cost RFID Tags. pp. 129–144.

Ferrag, M. A., L. A. Maglaras, H. Janicke, J. Jiang, and L. Shu (2017). Authentication Protocols for Internet of Things : A Comprehensive Survey. *2017*.

Flauzac, O., C. Gonzalez, and F. Nolot (2015). New security architecture for IoT network. *Procedia Computer Science 52*(1), 1028–1033.

From, C. (2005). ACCEPTED FROM OPEN CALL BLUETOOTH AND WI-FI WIRELESS PROTOCOLS :. (February), 12–26.

Guo, J., T. Peyrin, and A. Poschmann (2011). The photon family of lightweight hash functions. In P. Rogaway (Ed.), *Advances in Cryptology – CRYPTO 2011*, Berlin, Heidelberg, pp. 222–239. Springer Berlin Heidelberg.

Hassan, M., H. Mohammad, and R. Aref (2013). Attacks on Recent RFID Authentication Protocols. (August 2011).

He, D. and S. Zeadally (2015). An Analysis of RFID Authentication Schemes for Internet of Things in Healthcare Environment Using Elliptic Curve Cryptography. *2*(1), 72–83.

Janbabaei, S., H. Gharaee, and N. Mohammadzadeh (2016). Lightweight , Anonymous and Mutual Authentication in IoT Infrastructure. pp. 162–166.

Kenney, B. J. B. (2011). Dedicated Short-Range Communications (DSRC) Standards in the United States. *99*(7).

Khemissa, H. and D. Tandjaoui. A Novel Lightweight Authentication Scheme for heterogeneous Wireless Sensor Networks in the context of Internet of Things.

Landscape, C. and R. Technology. RPMA Technology for internet of things.

Latchman, H. A., S. Katar, L. Yonge, and S. Gavette (2013). *Homeplug AV and IEEE 1901: A Handbook for PLC Designers and Users* (1st ed.). Wiley-IEEE Press.

Lauridsen, M., H. Nguyen, B. Vejlgaard, P. Mogensen, M. Sørensen, and N.-i. Lora (2017). Coverage comparison of GPRS , NB-IoT , LoRa , and SigFox in a 7800 km 2 area. pp. 2–6.

Leander, G., C. Paar, A. Poschmann, K. Schramm, and G. Horst. New Lightweight DES Variants.

Lee, J.-y. and W.-c. Lin (2014). A Lightweight Authentication Protocol for Internet of Things. pp. 1–2.

Li, N. and D. Liu (2017). Lightweight Mutual Authentication for IoT and Its Applications. *14*(8).

Mackensen, E., M. Lai, and T. M. Wendt (2012). Bluetooth Low Energy (BLE) based wireless sensors. pp. 3–6.

Max, M. O. W. I. and K. Etemad (2009). M w max. (June), 82–83.

McKay, K. A., L. Bassham, M. S. Turan, and N. Mouha (2017). Report on lightweight cryptography.

Microsoft (2017). Reference Architecture. pp. 1–32.

Mohsen Nia, A. and N. K. Jha (2016). A Comprehensive Study of Security of Internet-of-Things. *IEEE Transactions on Emerging Topics in Computing PP*(99), 1–19.

Mundhenk, P., S. Steinhorst, M. Lukasiewycz, S. A. Fahmy, and S. Chakraborty (2015). Lightweight Authentication for Secure Automotive Networks. pp. 285–288.

N. Kushalnagar, G. Montenegro, C. S.

Rivest, R. L. 2 A Parameterized Family of Encryption Algorithms.

Sfar, A. R., E. Natalizio, Y. Challal, and Z. Chtourou (2017). A roadmap for security challenges in the internet of things. *Digital Communications and Networks*.

Sicari, S., A. Rizzardi, L. A. Grieco, and A. Coen-Porisini (2015). Security, privacy and trust in Internet of Things: The road ahead. *Computer Networks 76*, 146–164.

Summary, E., W. C. Latency, T. Rate, and S. L. Latencies. What is Network Latency and Why Does It Matter ? *O3bNetworks*, 1–13.

Sun, Y., H. Song, A. J. Jara, and R. Bie (2016). Internet of things and big data analytics for smart and connected communities. *IEEE Access 4*, 766–773.

Varma, V. K. (2012). Wireless Fidelity âĂŤ WiFi Wireless Fidelity âĂŤ WiFi. pp. 1–2.

Wu, Q., G. Ding, Y. Xu, S. Feng, Z. Du, J. Wang, and K. Long (2014). Cognitive Internet of Things: A New Paradigm Beyond Connection. *IEEE Internet of Things Journal 1*(2), 129–143.

Wu, W., S. Wu, L. Zhang, J. Zou, and L. Dong (2013). LHash : A Lightweight Hash Function (Full Version) âŃĘ. *Proceedings of Inscrypt 2013*.

Xu, L. D., W. He, and S. Li (2014, Nov). Internet of things in industries: A survey. *IEEE Transactions on Industrial Informatics 10*(4), 2233–2243.

Yassein, M. B., W. Mardini, and A. Khalil (2016). Smart Homes Automation using Z-wave Protocol.

Yi, S., Z. Qin, and Q. Li. Security and Privacy Issues of Fog Computing : A Survey. pp. 1–10.

You, W. and W. Learn (2014). The Internet of Things Reference Model. pp. 1–12.

You, W. and W. Learn (2015). Fog Computing and the Internet of Things : Extend the Cloud to Where the Things Are. pp. 1–6.

Chapter 2

![divider]

Mathematical Foundations

![divider]

CONTENTS

✓One must acknowledge with cryptography, no amount of violence will ever solve a math problem:

Jacob Appelbaum, Cypherpunks
Freedom and the Future of the Internet

2.1   Abstract

Cryptography without mathematics is something like earth without sun. Strong mathematical fundamentals help the cryptography researcher to understand and to develop new protocols. In this chapter we discuss basic mathematical aspects of cryptography starting from basic foundations like polynomial and group theory. Later on in this chapter, we have discuss some advanced concepts of elliptic curves and hash functions. Overall this chapter will help the reader understand the basic mathematics behind elliptic curve cryptography and other cryptography algorithms.

2.2 Elliptic Curve Cryptography

We have concluded in Chapter 1 that future cryptography will be a light weight cryptography. Elliptic curve cryptography attracted many researchers due to its smaller key size requirement compared to other cryptography algorithms. In this section, we will go through basic mathematical concepts that create a foundation for elliptic curve based cryptography. Some of the cryptographic schemes are designed in [Hou and Wang (2017), Luo et al. (2017), Zhang et al. (2015)].

2.2.1 Foundation

A small introduction to number theory will demonstrate clear game of numbers [Stallings (2010)].

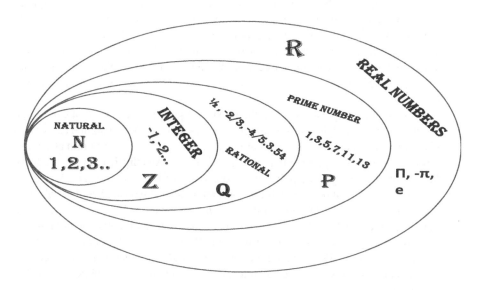

Figure 2.1: Numbers theory.

Figure 2.1 shows the distribution of number theory. Natural numbers are represented by $\mathbb{N}$ and contain numbers ≥ 1. Whole numbers are also natural numbers that include 0 also. Whole numbers can be represented by $\mathbb{W}$. Integer numbers range between $-N$ to $+N$ and can be represented by $\mathbb{I}$. Rational numbers are numbers that can be represented in the form of a/b, and they can be represented by $\mathbb{Q}$. Prime numbers are numbers that have GCD with any other number is 1. Prime numbers are the most important numbers in the world of cryptography. Prime numbers can be represented using $\mathbb{P}$. Real numbers are number line that contains all rational numbers, irrational numbers, fraction numbers and also transcendental numbers like $\Pi, 3\Pi, \gamma$. Complex numbers are numbers that can be represented by $c = a + ib$. Where $a, b \in \mathbb{R}$. "a" in complex number is the real part of numbers, "b" is imaginary part of complex number, and $i = \sqrt{-1}$. Complex numbers can be represented by $\mathbb{C}$.

$$\mathbb{N} \subset \mathbb{W} \subset \mathbb{I} \subset \mathbb{Q} \subset \mathbb{R} \subset \mathbb{C}$$

2.2.1.1 Basics of Polynomial Expressions

Algebraic expressions [Menezes et al. (1996)] are expressions that can be represented using variables, constants and some operations like addition, multiplication, substitutions, square root, etc. Examples of algebraic expressions are x, x^2, $x^3 + 2x + 5$, $3x^3 + 4x^2 + 5xy$. An algebraic expression in which exponents (or powers) of each terms are in the form of whole numbers then those algebraic expressions are called a polynomial expression. Examples of polynomial expressions are x^2, $x^3 + 2x + 5$. Algebraic expressions that are not polynomial expression examples are x^{-2}, $x^{-3} + 2x + 5$.

A polynomial expression where the maximum power in each term of expression 1 is called **Linear** polynomials. Examples are $ax + b, x + y$. A Polynomial expression where the maximum power in each term of expression 2 is called **Quadratic** polynomials. Examples of polynomials are $ax^2 + b, x^2 + y$. An **elliptic curve** is the curve that can be defined by using polynomial equations of form $y^2 = x^3 + ax + b$.

Co-coefficients of polynomials are constant numbers associated with variables of polynomials. Example: In polynomial $ax^2 + by + c$, where a, b, c are the exponents and x,y are the variables. Total number of terms associated with this polynomial is 3. Notation of polynomial with single variable is $P(x) = 8x^2 + 6x - 3$ or $P(u) = u^2 - u$ or $z(q) = q^3 + q + 3$. Polynomial functions with single variables can be defined as a $f(x)$. A polynomial with single variable x of maximum degree n can be represented as follows.

$$P(x) = a_0 + a_1 x + a_2 x^2 + \ldots\ldots\ldots + a_{n-1} x^{n-1} + a_n x^n$$

for all value of variable $x, n \geq 0$, and $a_0, a_1, \ldots\ldots, a_n$ are co-efficient.

In generalized way,

$$P(x) = \sum_{i=0}^{i=n} a_i x^i$$

Bit representation of polynomials:

Polynomials can be represented using bit representations. Examples: a polynomial $x^2 + x + 1$ can be represented using 111, while a polynomial $x^3 + 1$ can be represented using 1001. So if the coefficient of the variable is 0, we can say that the bit value of that term is 0.

Operations on Polynomials:
Let us take two polynomials $P(x) = ax^2 + bx + c$ and $Q(x) = px + q$

Polynomial additions:
$$R(x) = P(x) + Q(x) = ax^2 + (b+p)x + (c+q)$$

Polynomial Substitutions:
$$R(x) = P(x) - Q(x) = ax^2 + (b-p)x + (c-q)$$

Polynomial Multiplication:
$$R(x) = P(x) * Q(x) = apx^3 + aqx^2 + bpx^2 + bqx + cpx + cq$$

Polynomial Division:
Dividing polynomial $P(x) = x^3 - 2x^2 - 4$ with $Q(x) = x - 3$ will leave quotient $q(x) = x^2 + x + 3$ and remainder $r(x) = 5$. Polynomial division can be done via various methods like long division method. Polynomial $P(x) = Q(x)q(x) + r(x)$. if the $r(x) = 0$ means remainder is 0 than we can conclude that $Q(x)$ divides $P(x)$.

Some observations

1. Let us take polynomial P(x) with degree greater than equal to 1, let a be any real number then if $P(x)$ is divided by the linear polynomial $x - a$, remainder will be always $P(a)$.

2. If one of the polynomial is not defined over finite field, then division is not allowed. Example is polynomial defined over all integer or rational numbers.

3. A polynomial $P(x)$ is called as **irreducible polynomials or prime polynomials**, if it can not be expressed using product of two polynomials in the finite field with lower degree than $P(x)$.

4. A real number α is called root of quadratic equations $ax^2 + bx + c = 0, a \neq 0$, if $a\alpha^2 + b\alpha + c = 0$.

5. If $b^2 - 4ac \geq 0$ then the root of quadratic equation $ax^2 + bx + c$ are $-b \pm \sqrt{b^2 - 4ac}/2a$

6. If α and β are the roots of quadratic polynomial equation $P(x) = ax^2 + bx + c$ where $a \neq 0$ and $b \neq 0$. than $x - \alpha$ and $x - \beta$ are the factors of $P(x)$. $\alpha + \beta = -b/a$ and $\alpha * \beta = c/a$.

2.2.1.2 *Polynomial Curve fitting*

Polynomials [Menezes et al. (1996)] can be represented using points on graph. Connecting this point will generate a curve. That we can define as the curve for polynomials. Example of graphical representation in curves for the two polynomials $x^2 - x - 2$ and $x^3 - x^2 - 2$ is shown in Figure 2.2.

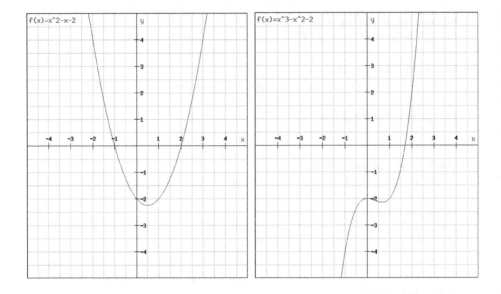

Figure 2.2: Polynomial curves example 1.

Following Figure 2.3 represents some of the elliptic curves by using polynomials like $y^2 = x^3 - x$, $y^2 = x^3 - 3x = 2$, $y^2 = x^3 - 2x + 1$ and circles like $x^2 + y^2 = 1$.

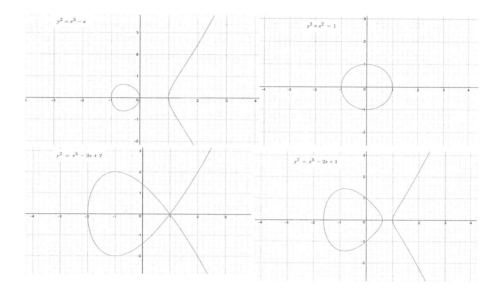

Figure 2.3: Polynomial curves example 2.

2.2.1.3 *Algebraic Set*

A clear understanding of finite field [Stallings (2010)] ensures an understanding of many cryptographic algorithms. To understand and implement modern cryptographic algorithms, the finite field plays an important role. The finite field is the foundation stone for elliptic curve cryptography. To understand the finite field, we will follow the hierarchy of learning algebraic theory. We will discuss algebraic theory, sets, groups, abelian group, ring, commutative ring, field, and finite field.

Every algebraic set has three important components:

1. Set of elements, set can be, examples: set of all integers($\mathbb{I}$), set of all prime numbers($\mathbb{I}_p$), set of all complex numbers($\mathbb{C}$).

2. Operations defined over set, examples: summation, multiplication, division, inverse.

3. Constant elements, examples: Identity elements like 0 for addition operation and 1 for multiplication operation.

Examples of algebra :

1. ($\mathbb{I}, +, 0$): Set of all integers that satisfies all set properties on operation "addition". Identity element for this set is 0. If a is a variable of set $\mathbb{I}$ then operation with identity element will return a.

2. $(\mathbb{R}, +, *, 1, 0)$: Set of all real numbers that satisfy all set properties on operation "addition" and "Multiplication". Identity element for addition operation is 0 and for multiplication operation is 1. Examples for any variable a of set, $a + 0 = a$ and $a * 1 = a$.

A general algebraic set can be defined as a $(\mathbb{S}, \Box, I)$, where $\mathbb{S}$ is set of elements, box represents operation and I represents identity elements.

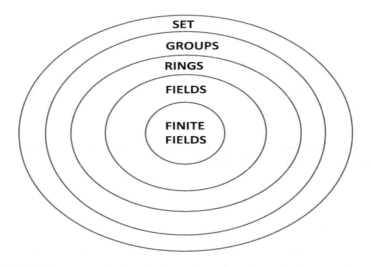

Figure 2.4: Algebraic Hierarchy.

Figure 2.4 shows algebraic hierarchy demonstrating that every finite field is a field. Every field is ring, every ring is group and every group is set.

Subset:

Let $\triangle$ and $\Box$ are two operations defined over set $\mathbb{S}$. Here $\triangle$ is binary operation and $\Box$ is unitary operation. Let $\mathbb{S}'$ be a subset of $\mathbb{S}$. Subset $\mathbb{S}'$ is closed with respect to $\triangle$ if for any two variable $a, b \in \mathbb{S}'$, $a \triangle b \in \mathbb{S}'$ and subset $\mathbb{S}'$ is closed with respect to $\Box$ if for any variable $a \in \mathbb{S}'$, $a \Box a \in \mathbb{S}'$.

Example: Let set $S = (\mathbb{I}, +, 0)$. And other set $S' = (X | 1 \leqslant X \leqslant 20)$. So we can say that $S' \subset S$. But set S' is not closed with respect to operation $+$. Cause if $a = 17, b = 12 \in S'$ then $a + b = 29 \notin S'$. But if considered max operation rather than $+$ operation then we can say S' is closed with respect to max operation.

Sub algebra:

Let $A = (S, \triangle, \square, I)$ and $A' = (S', \triangle', \square', I')$, where $S' \subset S, a\triangle' b = a\triangle b$ for all $a,b \in S', \square' a = \square a$, and $I' = I$. If A' is sub-algebra of a then it has same sign as A and follows the same operations as a. Example: Let $A = (I, +, 0)$ and $A' = (E, +, 0)$ where I is set of all integers and E is set of all even integers. we can say A' is sub algebra of a. It will satisfy all the operations and properties like a.

Properties:

Closure:

If $a, b \in S$ and operation defined over S is $\square$, then operation $a\square b = c$ is also in the set. Then we can say that set S satisfies the closure properties.

Commutative:

If $a, b \in S$ and operation defined over S is $\square$, Then operation $a\square b = b\square a$ is satisfied. Then we can say that set S satisfies the commutative property.

Associative:

If $a, b, c \in S$ and operation defined over S is $\square$, then operation $(a\square b)\square c = a\square(b\square c)$ is satisfied. Then we can say that set S satisfies the associative property.

Distributive::

If $a, b, c \in S$ and operations defined over S are $\square$ and $\triangle$, then operation $(a\square b)\triangle c = (a\triangle c)\square(b\triangle c)$ is satisfied. Then we can say that set S satisfies the distributive property.

Identity:

If $a \in S$, operation defined over S is $\square$ and identity $i \in \sim$ is defined, then operation $(a\square i) = a$ is satisfied. Then we can say that set S have an identity element.

Inverse:

If $a, b \in S$, operation defined over S is $\square$ and identity $i \in \sim$ is defined, then operation $a\square b = i$ is satisfied. Then we can say that b is inverse of a so we can say set S satisfies the inverse property.

Example: Real number set $(\mathbb{R}, +, *, -, 0, 1)$ and $a, b, c \in \mathbb{R}$, $+$ is addition operation and $*$ is multiplication operation than:

1. $a + b = b + a$ and $a * b = b * a$ (Commutative Property)

2. $a + (b + c) = (a + b) + c$ and $a * (b * c) = (a * b) * c$ (Associative Property)

3. $a * (b + c) = (a * b) + (a * c)$ (Distributive property)

4. $a + (-a) = 0$ and $a * (-a) = 1$ (Inverse property)

5. $a + 0 = a$ and $a * 1 = a$ (Identity property)

Various algebraic sets are distributed based on the set, operations defined on that set and the properties it satisfies.

2.2.1.4 *Groups*

■ **Semi group:**

Semi group is the algebra that satisfies the closure property and associative property for the operation defined on it, is called a semi group. Example of semi group is set $(\mathbb{I},+)$ and $(\mathbb{I},*)$, where for any three variable $a,b,c \in \mathbb{I}$, operation $a+b \in \mathbb{I}, a*b \in \mathbb{I}, a+(b+c) = (a+b)+c$ and $a*(b*c) = (a*b)*c$ is satisfied.

Remember: If $(S,\square)$ is semi group and $(T,\square)$ is sub algebra of $(S,\square)$ then $(T,\square)$ is also a semi group

■ **Monoid:**

Monoid is a semi group with existence of identity element. Examples of monoid are $(\mathbb{I},+,0)$ and $(\mathbb{I},*,1)$.

■ **Group:**

A group is denoted by $\mathbb{G}(S,\triangle)$. Group is a monoid that satisfies the inverse property also. So if $a,b,c,a^{-1} \in \mathbb{G}$, i is identity of group than:

1. $a\triangle b \in \mathbb{G}$ (Closure property)
2. $a\triangle(b\triangle c) = (a\triangle b)\triangle c$ (Associative property)
3. $a\triangle i = a$ (Identity element)
4. $a\triangle a^{-1} = i$ (Inverse element)

If the group has finite element then we can say it is **finite group** and if group has infinite elements as a set then we can say it is **Infinite group**.

■ **Abelian group:**

A group is called abelian group or commutative group if it satisfies the commutative property also. An abelian group is denoted by $\mathbb{G}(S,\triangle)$. So if $a,b,c,a^{-1} \in \mathbb{G}$, i is the identity of abelian group then

1. $a\triangle b \in \mathbb{G}$ (Closure property)
2. $a\triangle b = b\triangle a$ (Commutative)
3. $a\triangle(b\triangle c) = (a\triangle b)\triangle c$ (Associative property)
4. $a\triangle i = a$ (Identity element)
5. $a\triangle a^{-1} = i$ (Inverse element)

■ **Permutation group:**

Theorem 2.1

Permutation group is the Finite group

Proof 2.1   Let $S_n = \{1,2,3,........,n\}$ and $P_n = \{Set\ of\ all\ the\ permutations\ for\ elements\ 1,2,..n\}.\ Example\ S_3 = (1,2,3)\ than\ P_3 = ((1,2,3),(1,3,2),(2,1,3),(2,3,1),(3,1,2),(3,2,1)).\ Total\ number\ of\ sequences\ in\ the\ permutation\ set\ is\ 6\ which\ is\ finite\ number\ so\ cardinality(order* *of) of P_3 is 6.*

*Now consider that operation ◇ represents a binary composite operation on the permutation sequences. Binary composite operation performs re permutation of the second sequence according to first sequence. So if sequence $A, B \in P_n$ than operation $A \diamond B \in P_n$. Example : Let us take $A = (2,3,1)$ and $B = (1,3,2)$. than operation $A \diamond B = 3,2,1 \in P_3$. So we can say it satisfies the **closure property**.*

*Now let us take $A = (2,3,1), B = (1,3,2), C = (1,2,3)$. Then operation $A(2,3,1) \diamond (B(1,3,2) \diamond C(1,2,3)) = (3,2,1)$ and $(A = (2,3,1) \diamond B(1,3,2)) \diamond C(1,2,3) = (3,2,1)$. So we can consider that **associative property** is also satisfied.*

*Now if we perform operation ◇ with sequence $I = (1,2,3)$ on any sequence a will return the same sequence a. So sequence I represents **identity**.*

Now in the permutation set P_3, for every sequence a there exists an unique sequence α such that $A \diamond \alpha = I$. So α represents inverse sequence. Example : For sequence $(2,3,1)$, sequence $(3,1,2)$ is the inverse due to operation $(2,3,1) \diamond (3,1,2) = (1,2,3)$.

So we can say that permutation group P_n satisfies all the necessary properties for finite group [Menezes et al. (1996)].

Remember:

- If $\mathbb{G}(S, \triangle, \square, I)$ is group and $\mathbb{G}'(S', \triangle', \square', I')$ is sub-algebra of $\mathbb{G}$ then $\mathbb{G}'$ is also called as a subgroup of $\mathbb{G}$. So "a sub-algebra of group is a group".

- Order of the group can be represented by $|\mathbb{G}|$. $|\mathbb{G}|$ is total number of elements in the set defined over group. Order of finite group is the finite integer number.

- A group is a monoid in which every element is invertible.

■ **Cyclic Group** A group $\mathbb{G}$ is cyclic if it contains a generator ξ. Generator is the element of group through which we can generate remaining elements of the group [Stallings (2010)]. Example: if $a = \xi \in \mathbb{G}$. Then for every element $b_i \in \mathbb{G}$, we can represent it as a $a^j = b_i$. Example: Consider the group $\mathbb{G}_{11} = \{1,2,3,4,5,6,7,,8,9,10\}$ defined over operation modulo 11. Then, for every $i \leq 10$, operation $2^i \bmod 11$ will return $(2,4,8,5,10,9,7,3,6,1)$, which is a set of all elements defined in group. So we can say 2 is the generator of group $\mathbb{G}_{11}$ defined over operation modulo 11. A group that contains only one generator is called a cyclic group. Every cyclic group is a commutative group, but it is not necessary that every commutative group is cyclic. Cyclic groups can be finite or infinite. Elements with the maximum orders are called a primitive element or generator.

2.2.1.5 Rings

A ring, R, can be represented as a $\mathbb{R}\{S,+,\times\}$. A ring R is an abelian group with respect to operation $+$ and satisfies closure, commutative, associative property and contains identity element and inverse element for operation $+$. With respect to it, ring has additional property for new operation $\times$.

- let $a,b \in R$ then operation $a \times b \in R$. (Closure over $\times$)

- let $a,b,c \in R$ then operation $a \times (b \times c) = (a \times b) \times c$. (Associative over $\times$)

- let $a,b,c \in R$ then operation $a \times (b+c) = (a \times b) + (a \times c)$ and $(a+b) \times c = (a \times c) + (a \times b)$. (Distributive over $\times$)

- let $a,b \in R$ than $a \times b = b \times a$. (Commutative over $\times$)

A ring is an integral domain if it follows all the above said properties and also the following two properties.

- let $1 \in R$ is the identity element than for $a \in R$, operation $a \times 1 = 1 \times a = a$. (Multiplicative identity element).

- let $a,b \in R$ then if $a \times b - 0$, than either $a - 0$ and $b = 0$. (Absence of zero divisor)

Example : Let $\mathbb{R}$ is the set of real numbers, $\mathbb{C}$ is the set of complex numbers and $\mathbb{Q}$ is the set of rational numbers than $(\mathbb{R},+,\times),(\mathbb{C},+,\times),(\mathbb{Q},+,\times)$ are the rings. Set of all the integers, set of all the even integers are the examples of ring.

2.2.1.6 Fields and Finite fields

1. **Field:** A field F is the ring with the additional operation multiplicative inverse. So an algebra $F(S,+,\times)$ is the field with two binary operation $+$ and $\times$ if,

   - F over operation $+$ is abelian group.
   - For every element $a \in F$, except 0, F is abelian group. So there exists another element a^{-1} such that $a \times a^{-1} = a^{-1}$ times $a = 1$.
   - For every element $a,b \in F$ then $a/b = a \times b^{-1}$.
   - For every element $a,b,c \in F$ then $a \times (b+c) = a \times b + a \times c$ and $(a+b) \times c = a \times c + b \times c$.

   Example: A set of all the real numbers, rational numbers, complex numbers is field. While a set of all the integers is not a field [Menezes et al. (1996)] because there is no existence of multiplicative inverse for elements in integers except 1 and -1.

2. **Finite field:** In the world of cryptography, we are not interested in the infinite field. Finite field can be represented as a $\mathbb{GF}$. Finite field is also called a galois field. A field whose order (number of elements) is in the power of prime

element *P*. So a finite field has P^m elements, where P is the prime number and m is the positive integer number. A finite field with 13 element exists (13^1), A finite field with 81 elements exists$(3^4 = 81)$. There two types of finite field that exists:

(a) If $m = 1$, than finite field $\mathbb{GF}$ is $\mathbb{GF}(P)$ and it is called as prime field.

(b) if $m > 1$, than finite field $\mathbb{GF}$ is $\mathbb{GF}(p^m)$ and is called as extension field. In the cryptography, we mostly use binary extension field $\mathbb{GF}2^m$ which defines the input key size.

The following figure shows a summary for famous algebraic and their operations.

OPERATION ⟹ ALGEBRA ⬇	Associativity +	Associativity *	Closure +	Closure *	Commutative +	Commutative *	Inverse +	Inverse *	identity +	identity *	No zero Division
GROUP	YES	NO	YES	NO	NO	NO	YES	NO	YES	NO	NO
ABELIAN GROUP	YES	NO	YES	NO	YES	NO	YES	NO	YES	NO	NO
RING	YES	YES	YES	YES	NO	YES	YES	NO	YES	YES	NO
ABELIAN RING	YES	YES	YES	YES	YES	YES	YES	NO	YES	YES	NO
INTEGRAL DOMAIN	YES	YES	YES	YES	YES	YES	YES	NO	YES	YES	YES
FIELD	YES	YES	YES	YES	YES	YES	YES	YES	YES	YES	YES

Figure 2.5: Algebraic operation summary.

In Figure 2.5, $+$ represents additive and $*$ represents multiplicative operations.

2.2.1.7 GCD, Modular arithmetic, Prime numbers

■ **Divisors and divisibility** we can say a variable a divides b if we can represent a = m*b, where a, b, and m are the integer numbers. We can represent b divides a by b/a if remainder is 0.

So in general division, we can represent as follows

$a = m * b + r$, where a and b are positive integers, and *m* is quotient and *r* is remainder of the division operation. Value of remainder will be always between 0 and m. Example: $35 = 2 * 12 + 11$.

■ **GCD**

- Greatest common divisor of a and b is represented as GCD(a,b). GCD of two number a and b represents the largest number that divides both the numbers a and b.

- $GCD(a,b)$ = maximum[m, divisors of a and b]. Example $GCD(12,15) = 3$.

- if $GCD(a,b) = 1$ then we can say that a and b are relatively prime numbers to each other.

- $GCD(a,a) = a$ and $GCD(a,0) = a$.

- If b/a than $GCD(a,b) = b$.

■ **Euclidean Algorithms**

Euclidean algorithms can be used to find the GCD between two large number [Menezes et al. (1996)]. GCD can be found by using prime factorization operation but this can work only for small size of integers. So Euclidean algorithm provides an other way to find the GCD between two numbers.

$$\text{GCD}(a_0,a_1) = \text{GCD}(a_0 \bmod a_1, a_1) = \text{GCD}(a_1, a_0 \bmod a_1)$$

Above computations will be iterated until we get $GCD(a_1,0) = a_1$.

Example: GCD(973,301) using Euclidean algorithm can be found in the following steps, let $a_0 = 973$ and $a_1 = 301$.

- GCD(973,301) = GCD(70,301) = GCD(301,70) = GCD(21,70)
- GCD(21,70) = GCD(70,21) = GCD(7,21)= GCD(21,7) = GCD(7,0)= 7

■ **Extended Euclidean Algorithms**

Extended Euclidean algorithms can help in the calculation of finite field and some advanced encryption algorithms. For any two integers a_0 and a_1, GCD is g, and p,q are two additional integers of opposite sign as a coefficient, which will satisfy the Extended Euclidean equation. Extended Euclidean algorithm can be defined as follow:

$$\text{GCD}(a_0, a_1) = pa_0 + qa_1 = g$$

Computation of Extended Euclidean algorithm computed per following steps:

- $\text{GCD}(a_0,a_1) \Rightarrow a_0 = pa_1 + a_2 \Rightarrow a_2 = p_1 a_0 + q_{(}1)a_1$
- $\text{GCD}(a_1,a_2) \Rightarrow a_1 = pa_2 + a_3 \Rightarrow a_3 = p_2 a_1 + q_{(}2)a_2$

■ and ....

■ $GCD(a_{n-1}, a_{n-2}) \Rightarrow a_{n-1} = pa_{n-2} + a_n \Rightarrow a_n = p_n a_{n-1} + q_(n-2)a_1$

Example: Let us compute GCD($973 = a_0, 301 = a_1$) using Extended Euclidean algorithm.

■ $GCD(973, 301) = p973 + q301 = 7$ so we want to compute the value of p and q.

■ $973 = 3 * 301 + 70 \Rightarrow 70 = 1.973 + (-3) * 301$

■ $301 = 4 * 70 + 21 \Rightarrow 21 = 1 * 301 + (-4) * 70 = 1 * 301 + (-4) * (1 * 973 + (-3) * 301) = (-4) * 973 + (13) * 301$

■ $70 = 3 * 21 + 7 \Rightarrow 7 = 1 * 70 + (-3) * 21 = 1 * (1.973 + (-3) * 301) + (-3) * ((-4) * 973 + (13) * 301) = 13 * 973 + (-42) * 301$

■ 7 can not be further divided in the required form so we can say our required GCD is 7 and value of $p = 13$ and $q = (-42)$.

■ **Modular arithmetic:** The basic aim of modular arithmetic is to do the computation in the finite set.

■ If a is an integer and p is another positive integer, then the remainder of operation a divides p is the answer of a mod p. Example: 17 mod 3 = 2, if the value of a is lesser than p then the remainder will be a it self.

■ Two integers a and b are congruent modulo p it means a mod p = b mod p and it can be represented by a ≡ b mod p.

Example: let us take a = 7 and p = 2 than the possible value of b in the congruence relation can be defined as follow:

■ $7 \equiv 1 \bmod 2$
■ $7 \equiv 3 \bmod 2$
■ $7 \equiv 5 \bmod 2$
■ $7 \equiv 7 \bmod 2$
■ $7 \equiv -1 \bmod 2$
■ $7 \equiv -3 \bmod 2$

so left side values of mod operation if we write as a set $\{.... -3, -1, 1, 3, 5, 7....\}$ creates an equivalence class and all the members of equivalence class are relatively prime to each other.

■ If a ≡ b mod p than p divides (a-b) or b-a. So if p divides a-b and answer is some constant integer q than a = b + pq.

■ If a ≡ b mod p than b ≡ a mod p.

■ If a ≡ b mod p, b ≡ c mod p than a ≡ c mod p.

■ If GCD(a,p) = 1 then we can say inverse exists for the a and if GCD(a,p) ≠ 1 then inverse does not exists for a.

■ Example : if a= 3^8 and p = 7 find out the answer of a mod p.

- - Calculation way 1 : calculate $3^8 = 6567$ and calculate 6567 mod 7 = 2.
  - Calculation way 2 : Divide 3^8 mod 7 as $(3^4 * 3^4)$ mod 7 = (4*4) mod 7 = (16 mod 7) = 2.
- a is an additive inverse of b if $(a+b) \equiv 0$ mod p. For every integer there exists an additive inverse.
- a is an multiplicative inverse of b of $(a*b) \equiv 1$ mod p. a has a multiplicative inverse in (mod p) if a is relatively prime to p. It is not necessary that multiplicative inverse should exist for all integers.
  Example: if p = 8 than for $\mathbb{Z}_p$ = (0,1,2,3,4,5,6,7)
  - Elements : 0 , 1 , 2, 3 , 4 , 5 , 6 , 7
  - Additive inverse : 0 , 7 , 6 , 5 , 4 , 3 , 2 , 1
  - Multiplicative inverse : - , 1 , - , 3 , - , 5 , - , 7

  So we can observe that the multiplicative inverse exists only for the numbers that are relatively prime number to mod p.
- $a/b \equiv a*a^{-1}$ mod p.
- (mod p) operation can be denoted by $\mathbb{Z}_p$ where $\mathbb{Z}_p$ ={0,1,2,3,......p-1} so $\mathbb{Z}_p$ is the set of remainders in the arithmetic (mod p).
- Properties of Z_p, let a,b,c $\in Z_{(p)}$
  - (a+b) mod p = (b+a) mod p and (a*b) mod p = (b*a) mod p.
  - (a+b) + c mod p = a+(b+c) mod p and (a*b) * c mod p = a*(b*c) mod p.
  - a*(b+c) mod p = (a*b) mod p + (a*c) mod p.
  - (0+a) mod p = (a+0) mod p and (1*a) mod p = (a*1) mod p.
  - a + (-a) = 0 mod p, both a,-a $\in \mathbb{Z}_p$.
- (a mod p) + (b mod p) = (a+b) mod p, (a mod p) - (b mod p) = (a-b) mod p,(a mod p) * (b mod p) = (a*b) mod p.
- $(a+b) \equiv (a+c)$ mod p $\Rightarrow b \equiv c$ (mod p) but $(a*b) \equiv (a*c)$ mod p $\not\Rightarrow b \equiv$ c (mod p) unless a and p are relatively prime to each other.
- if value of a is multiple of p like (p,2p,3p,......) than $a \equiv 0$ mod p.
- if value of a is p-1 or multiple of p-1 than $a \equiv -1$ mod p.
- If $a \equiv b$ mod p & $c \equiv d$ mod p than $-a \equiv$ mod p, $(a+c) \equiv (b+d)$ mod p and $(ac) \equiv (bd)$ mod p.

- **Prime Numbers**

  - Prime numbers play an important role in cryptography algorithms [Stallings (2010)].
  - A number *p* is prime number if the factoring operation on *p* will give ± 1 or $\pm p$. smallest prime number is 2.

■ any number $a > 1$ can be represented as a

$$a = p_1{}^{x_1} + p_2{}^{x_2} + \ldots\ldots\ldots\ldots p_t{}^{x_t} \text{ where } p_1 < p_2 < p_3 \ldots\ldots\ldots < p_t$$

where p_i is the prime number.

So in general $a = \prod_{p \in P} P^{a_p}$ where $a_p \geq 0$.

■ Example : $17 = 1^1 + 17^1$, $3600 = 2^4 + 3^2 + 5^2$

■ Euler's totient function

■ Suppose $a \geq 1 \& p \geq 2$ are the integers then if $GCD(a, p) = 1$ then we can say that a and p are relatively prime to each other.

■ The number of integer in $\mathbb{Z}_{,}$ where $(p > 1)$ that are relatively prime to p does not exceed p is denoted by $\phi(p)$ called as Euler's totient function or phi function.

■ $\phi(1) = 1$.

■ If $p = 26$ than relative prime numbers to 26 = (1, 3, 5, 7, 9, 11, 15, 17, 19, 21, 23, 25) so $\phi(26) = 12$. Other example if $p = 35$ than all the positive integer less than 35 and relatively prime to 35, are (1, 2, 3, 4, 6, 8, 9, 11, 12, 13, 16, 17, 18, 19, 22, 23, 24, 26, 27, 29, 31, 32, 33, 34) so $\phi(35) = 24$.

■ if p is prime number than $\phi(p) = p - 1$.

■ $\phi(pq) = \phi(p) * \phi(q)$, Example $\phi(77) = \phi(11 * 7) = 10 * 6 = 60$.

■ *Theorem 2.2*

Let a is having following canonical factorization
$a = p_1{}^{x_1} * p_2{}^{x_2} * p_3{}^{x_4} * \ldots\ldots * p_n{}^{x_n}$ *where p is prime number and x is integer number then* $\phi(a) = \prod_{i=1}^{n}(p_i{}^{e_i} - p_{i-1}{}^{e_i-1})$ *where* $\prod$ *is the product operation.*

Example: $a = 240$ then finding $\phi(240)$.

■ Factorizing $240 = 2^4 * 3 * 5$, so $p_1 = 2, p_2 = 3, p_3 = 5, e_1 = 4, e_2 = 1, e_3 = 1$ and $n = 3$(total numbers of p).

■ so according to theorem $(2^4 - 2^3)(3^1 - 3^1)(5^1 - 5^0)$

■ $8 * 2 * 4 = 64$

■ So $\phi(240) = 64$ means total number of relative primes to 240 are 64.

■ For two different prime numbers p and q if $n = p * q$ than $\phi(n) = (p - 1) * (q - 1)$

■ For every a and p that are relatively prime, $a^{\phi(p)} \equiv 1 (mod\, p)$

■ For every positive integer a and p , $a^{\phi(p)+1} \equiv 1 (mod\, p)$

■ Fermat's theorem

■ If p is prime & a is positive integer not divisible by p then $a^{p-1} \equiv 1$ (mod p)

■ If p is prime & a is positive integer then $a^p \equiv a(\text{mod } p)$

2.2.1.8   Basics of discrete logarithms

■ **Log in normal arithmetic(not in modulo arithmetic):**
For any 3 integer number a, b, i if $b = a^i$ then we can say that $i = log_a b$

■ **Log in modulo arithmetic):**
For any integers a,b,i integers and p is prime number then if $b = a^i mod p$ than $i = dlog_{a,p} b$

Calculating exponent i:

■ A unique exponent i can be found only if a is primitive root of odd prime number p.

■ If a is primitive root of p then $(a^1 mod p, a^2 mod p, a^3 mod p, \ldots \ldots, a^{p-1} mod p)$ will return unique answers.

■ Not all the integers have primitive roots, primitive roots only exist for the integers with the form of $2, 4, p^\alpha, sp^\alpha$

■ Examples: Finding the primitive root for mod 7,

a	a^1 Mod 7	a^2 Mod 7	a^3 Mod 7	a^4 Mod 7	a^5 Mod 7	a^6 Mod 7
1	1	1	1	1	1	1
2	2	2	1	2	4	1
3	3	2	6	4	5	1
4	4	2	1	4	2	1
5	5	4	6	2	3	1
6	6	1	6	1	6	1

Figure 2.6: Primitive root.

For mod 7, primitive roots are 3 and 5 as shown in Figure 2.6. The following Figure 2.7 shows value of exponent for the base 3 and 5 for mod 7.

$i = \text{dlog}_{a,p} b \ \Rightarrow \ b = a^i \bmod p$					
p = 7					
Base a = 3					
b	1	2	3	4	5	6
i	6	2	1	4	5	3
Base a = 5					
b	1	2	3	4	5	6
i	6	4	5	2	1	3

Figure 2.7: Exponent.

■ So the discrete logarithm problem can be defined as follow.

"Discrete logarithm problem [Menezes et al. (1996)] in the cyclic group $\mathbb{Z}_p^*$, where p is prime so the order of group is p-1. Primitive root $a \in \mathbb{Z}_p^*$ and other element $b \in \mathbb{Z}_p^*$ then the discrete logarithm problem over the multiplication operation for to find exponent i which is $0 \le i \le p-1$, such that $b = a^i \bmod p$. Finding the exponent i becomes very difficult for the larger value of prime p".

■ **One way function** :

A function $f()$ is a one way function if $y = f(x)$ is easy to compute in polynomial time but reverse operation on the function means $x = f^{-1}(x)$ is very hard to compute in polynomial time. Famous examples of one way functions used in cryptography are factorizing multiplication of two large primes used in RSA and discrete logarithm problems used in Diffie-Hellman key exchange algorithms.

2.2.1.9 *Polynomial arithmetic on prime field and extension field*

We have seen in the previous section about basics of polynomial and its operation. In this section we will study polynomial division, polynomial operations in prime field $\mathbb{GF}(p)$ and in the extension field $\mathbb{GF}(2^m)$. We will study how modular arithmetic plays a role in the computation of polynomial [Menezes et al. (1996)].

We can divide polynomial $3x^3 - 5x^2 + 10x + 3$ by $2x + 1$ using long division method as follow :

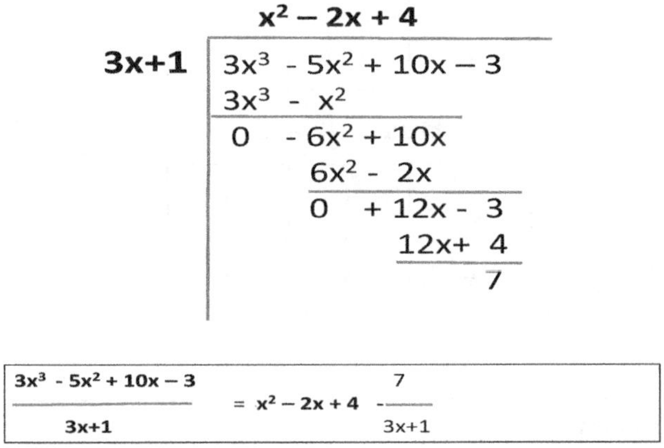

$$
\begin{array}{r}
\mathbf{x^2 - 2x + 4} \\
\hline
\end{array}
$$

$$
\mathbf{3x+1} \,\big|\,
\begin{array}{l}
3x^3 - 5x^2 + 10x - 3 \\
\underline{3x^3 - x^2} \\
\quad 0 \;\; -6x^2 + 10x \\
\quad\quad \underline{6x^2 - 2x} \\
\quad\quad\quad 0 \;\; + 12x - 3 \\
\quad\quad\quad\quad \underline{12x + 4} \\
\quad\quad\quad\quad\quad\quad \cdot 7
\end{array}
$$

$$
\frac{3x^3 - 5x^2 + 10x - 3}{3x+1} = x^2 - 2x + 4 \; - \frac{7}{3x+1}
$$

Figure 2.8: Polynomial Division.

Figure 2.8 shows how to perform polynomial decision operations.

■ **Irreducible polynomial and Prime polynomial:** A polynomial $p(x)$ in the field F is called irreducible polynomial if it can not be represented in the multiplication of two polynomials with lesser degree than $p(x)$ in the same field. For every finite field of $\mathbb{GF}(2^m)$ there are several irreducible polynomials. For every degree there is a list of irreducible polynomials. Irreducible polynomials are the prime numbers in the world of polynomials [Menezes et al. (1996)]. Example: For $\mathbb{GF}(2^m)$, set of irreducible polynomial of degree from $(1, 2, 3, \ldots, m)$.

    ■ Degree 1 : $x + 1, x$

    ■ Degree 2 : $x^2 + x + 1$

    ■ Degree 3 : $x^3 + x + 1, x^3 + x^2 + 1$

■ Degree 4 : $x^4 + x + 1, x^4 + x^3 + x^2 + x + 1, x^4 + x^3 + 1$

Irreducible polynomial for AES algorithm where key size is 256 bit($2^8 bit$) = $x^8 + x^4 + x^3 + x + 1$. We will signify irreducible polynomial by $p(x)$. Major operation in the field are addition, subtraction, multiplication, division or inversion.

■ **Prime field $\mathbb{GF}(p)$ arithmetic:**
let $a, a^{-1}, b, c, d, e, f \in \mathbb{GF}(p) = \{0, 1, 2, \ldots\ldots p - 1\}$

   ■ a+b $\equiv$ c mod p

   ■ a-b $\equiv$ d mod p

   ■ a*b $\equiv$ e mod p

   ■ $a * a^{-1} \equiv 1$ mod p.

■ **Extension field $\mathbb{GF}(2^m)$ arithmetic:**

   ■ **Element representation in the extension field:** For coefficient a and power m, a extension field $\mathbb{GF}(2^m)$ can be defined as follow:

$$f(x) \in \mathbb{GF}(2^m) = a_0 + a_1 x^1 + a_2 x^2 + \ldots\ldots + a_{m-2} x^{m-2} + a_{m-1} x^{m-1}$$
$$\text{for } a_i \in \mathbb{GF}(2^m)$$

   Example : Let us take $\mathbb{GF}(2^4)$ so $\mathbb{GF}(16)$ will have four terms and it can be represented as $f(x) = a_3 x^3 + a_2 x^2 + a_1 x^1 + a_0$, values of $a_1, a_2, a_3, a_4 \in \{0, 1\}$ so possible polynomials in $\mathbb{GF}(2^4)$ is as shown in following Table 2.1:

Table 2.1: Binary representation of polynomials.

c_3	c_2	c_1	c_0	**polynomial**
0	0	0	0	0
0	0	0	1	1
0	0	1	0	x
0	0	1	1	x+1
0	1	0	0	x^2
0	1	0	1	$x^2 + 1$
0	1	1	0	$x^2 + x$
0	1	1	1	$x^2 + x + 1$
1	0	0	0	x^3
1	0	0	1	$x^3 + 1$
1	0	1	0	$x^3 + x$
1	0	1	1	$x^3 + x + 1$
1	1	0	0	$x^3 + x^2$
1	1	0	1	$x^3 + x^2 + 1$
1	1	1	0	$x^3 + x^2 + x$
1	1	1	1	$x^3 + x^2 + x + 1$

■ **Addition and subtraction in the extension field of** $\mathrm{GF}(2^m)$**:**
Let $a(x), b(x) \in \mathrm{GF}(2^m)$ a polynomials, then the addition operation of polynomials in the $\mathrm{GF}(2^m)$ can be defined as follow :

$$c(x) = a(x) + b(x) = \Sigma_{i=0}^{m-1} c_i x^i \text{ where } c_i(x) \equiv (a_i(x) + b_i(x)) \bmod 2$$

Example: Let field $\mathrm{GF}(2^4)$ and a(x)$= x^3 + x^2 + 1$ and b(x)$= x^3 + x^2 + x + 1$ than c(x) = x. Subtraction is similar to the addition operation in the $\mathrm{GF}(2^m)$.

■ **Multiplication in the extension field of** $\mathrm{GF}(2^m)$**:**
Multiplication operation in the $\mathrm{GF}(2^m)$ is similar to the regular polynomial multiplication but in the $\mathrm{GF}(2^m)$ if multiplication polynomial is not within $\mathrm{GF}(2^m)$ then it will be reduced using irreducible polynomial.
Example: Let field $\mathrm{GF}(2^4)$ and $a(x) = x^3 + 1$ and $b(x) = x + 1$ than $c(x) = (a(x) * b(x)) \bmod 2 = (x^4 + x^3 + x + 1)$. We can notice that $c(x) \notin \mathrm{GF}(2^4)$.
Now let us consider c(x) as $c'(x)$ and take irreducible polynomial p(x) of any degree but $p(x) \in \mathrm{GF}(2^4)$. For example $p(x) = x$ we have taken than $c'(x)$ mod p(x) will give remainder $1 \in \mathrm{GF}(2^4)$. So we can say that multiplication answer is 1 and we can observe that computation depends on the irreducible polynomial which we choose for the computation.

■ **Inverse in the extension field of** $\mathrm{GF}(2^m)$**:**
Let us take $a(x), a^{-1}(x) \in \mathrm{GF}(2^m)$ and $p(x)$ is irreducible polynomial than the inverse operation in $\mathrm{GF}(2^m)$ can be defined as follow.

$$a(x) * a^{-1}(x) \equiv 1 \bmod p(x).$$

2.3 Diffie-Hellman key Exchange and Elgamal scheme

Public key cryptography is a famous type of cryptography in which two keys are used for security. One key is the public key, which is publicly available to all the participants. And the other is a private key, which is available only to sender or receiver. The private key is not shared with any-one. Public key cryptography is mainly used for encryption and authentication. Encryption provides confidentiality of data. Authentication ensures that both communicating parties are certain about each others identity. The following figures show confidentiality and authentication in general cryptography.

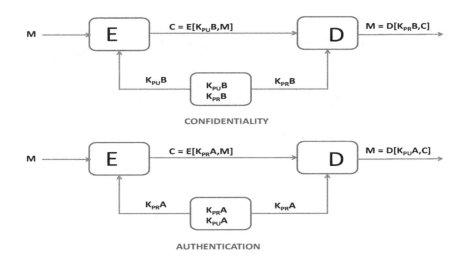

Figure 2.9: Confidentiality and Authentication.

Basic operation of confidentiality and authentication is shown in Figure 2.9. Confidentiality can be achieved using public key of receiver for encryption, and authentication can be achieved using private key of sender for encryption.

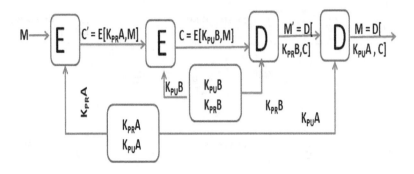

CONFIDENTIALITY AND AUTHENTICATION

Figure 2.10: Complete Secrecy.

Figure 2.10 shows complete public key cryptography, which provides secrecy and confidentiality at a same time.

2.3.1 *Diffie-Hellman key exchange*

Diffie-Hellman key exchange [Stallings (2010)] was proposed by Whitfield Diffie and Martin Hellman in 1976. This was the first public key cryptography algorithm which was proposed for session key establishment with out direct exchange.

■ Basic principle behind Diffie-Hellman key exchange is discrete logarithm problem, cyclic group and prime number.

■ $K \equiv \alpha^{a*b} mod\, p = K \equiv \alpha^{b*a} mod\, p$

■ some of the notations for the algorithms we will use are : $K_{PU_A}=$ public key of alice, $K_{PU_B}=$ public key of bob, $K_{PR_A} \in \{2,3,4.....p-2\}$ is the private key of Alice. $K_{PR_B} \in \{2,3,4.....p-2\}$ is the private key of bob, base $\alpha \in \{2,3,4.....p-2\}$, p is large prime number and $K_{A,B}$.

■ In the initialization phase of Diffie-Hellman key exchange, the algorithm value of α and p becomes public. So communicating parties can use it for computing.

- Alice chooses $K_{PR_A} \in \{2,3,4.....p-2\}$
- Alice computes $K_{PU_A} = \alpha^{K_{PR_A}} \bmod p$

- Bob chooses $K_{PR_B} \in \{2,3,4.....p-2\}$
- Bob computes $K_{PU_B} = \alpha^{K_{PR_B}} \bmod p$

- Alice shares K_{PU_A} publicly with bob
- Bob shares K_{PU_B} publicly with Alice

- Alice Computes $K_{A,B} = K_{PU_B}^{K_{PR_A}} \bmod p$
- Bob Computes $K_{A,B} = K_{PU_A}^{K_{PR_B}} \bmod p$
- $K_{A,B}$ computed by Alice and $K_{A,B}$ computed by bob will be same. Because we can easily under stand that $K_{A,B} = \alpha^{K_{PR_A}*K_{PR_B}} \bmod p$

Diffie-Hellman key exchange algorithm is vulnerable if the attacker can solve the discrete logarithm in the polynomial time. Example of Diffie-Hellman key exchange algorithm: Let prime number $p = 353$ and a primitive root of p, $\alpha = 3$. Now Alice chooses private key $K_{PR_A} = 97$ and Bob chooses private key $K_{PR_B} = 233$. Now Alice computes $K_{PU_A} = \alpha^{K_{PR_A}} \bmod p \Rightarrow 3^{97} \bmod 353 = 40$ and Bob computes $K_{PU_B} = \alpha^{K_{PR_B}} \bmod p \Rightarrow 3^{233} \bmod 353 = 248$.

So now Alice and Bob share K_{PU_A} and K_{PU_B}, now Alice computes $K_{A,B} = K_{PU_B}{}^{K_{PR_A}} \bmod 353 = 248^{97} \bmod 353 = 160$. Similarly Bob computes $K_{A,B} = K_{PU_A}{}^{K_{PR_B}} \bmod 353 = 40^{233} \bmod 353 = 160$.

Attacks on Diffie-Hellman: There are famous attacks like man-in-the middle attack, brute-force attack, Index calculus attack and square root attacks.

2.3.2 Elgamal crypto system

Elgamal, in 1984, introduced one new protocol that was a close relative to the Diffie-Hellman key exchange and was on the same computation problem of the discrete logarithm problem [Stallings (2010)].

- Bob chooses prime number p and primitive element α.
- Bob chooses $K_{PR_B} \in \{2,3,4,....p-2\}$.
- $K_{PU_B} = \alpha^{K_{PR_B}} \bmod p$
- Bob shares publicly K_{PU_B}, p, α.

- Alice has plain text $M \in \{2,3,4,....p-2\}$

- Alice chooses $K_{PR_A} \in \{2,3,4,....p-2\}$.

- Alice computes $K_T = \alpha^{K_{PR_A}} \bmod p$.

- Alice computes $K_M = K_{PU_B}^{K_{PR_A}} \bmod p$.

- Alice prepares cipher text $C_M = K_M * M \bmod p$.

- Alice sends cipher text and temporary key (K_T, C_M).

- Bob receives (K_T, C_M).

- Bob computes $K_M = K_T^{K_{PR_B}} \bmod p$.

- Bob computes $M = (C_M * K_T^{-1}) \bmod p$.

Elgamal crypto system has one advantage over Diffie-Hellman key exchange: that the value of K_T will be different for each message communication but at the same time it increases computation over head. So it may not be possible to use in lightweight cryptography.

2.4 Elliptic Curve Cryptography Operations

Elliptic curve [Stallings (2010), Menezes et al. (1996)] attracted cryptographers due to its smaller key size compared to other algorithms and provided equal security levels. The following figure shows comparison table between the most famous cryptography algorithms RSA, Diffie-Hellman and ECC. ECC is the prominent cryptographic direction in the world of light weight cryptography.

Comparison between Diffie-Hellman, RSA and ECC is shown in Figure 1.11. Diffie-Hellman algorithms use multiplication to show exponentiation. Diffie-Hellman algorithms use $\alpha^k \bmod p = \{\alpha * \alpha * ..... * \alpha\}$ for k times, while elliptic curve performs multiplication operation using repeated additions. $\alpha * k = \{\alpha + \alpha + \alpha + ... + \alpha\}$ for k times. So this multiplication operation reduces complexity over exponentiation operation.

Symmetric Key Algorithms	Diffie-Hellman and Digital signature	RSA	ECC
80	K_{PUB} : 1024 K_{PR} : 160	1024	160-223
112	K_{PUB} : 2048 K_{PR} : 224	2048	224-255
128	K_{PUB} : 3072 K_{PR} : 256	3072	256-383
192	K_{PUB} : 7680 K_{PR} : 384	7680	384-511
256	K_{PUB} : 15,360 K_{PR} : 512	15360	512

Figure 2.11: Comparison of Diffie-Hellman, RSA, and ECC.

Elliptic curve can be defined on the variables and coefficient associated with the variables. A generalized elliptic curve can be defined as a:

$$y^2 = x^3 + ax + b \tag{2.1}$$

In the equation a and b are real numbers. Elliptic curve is considered a cubic polynomial. We can represent elliptic curve as E(a,b), which consists of all the points (x,y) that satisfy 2.1. Elliptic curve variable can be defined as :

$$E(a,b) = \forall_{y^2=x^3+ax+b}(x,y) \bigcup \{O\} \tag{2.2}$$

In the above equation one new point is {0}, which is point of infinity in the elliptic curve, which works as a additive identity in the elliptic curve operation. Examples are $y^2 = x^3 - x$ where a = -1 and b = 0, $y^2 = x^3 + x + 1$ where a=1 and b =1. Value of a and b must follow equation $4a^3 + 27b^2 \neq 0$. The following figure shows elliptic curve for $x^3 + x + 1$.

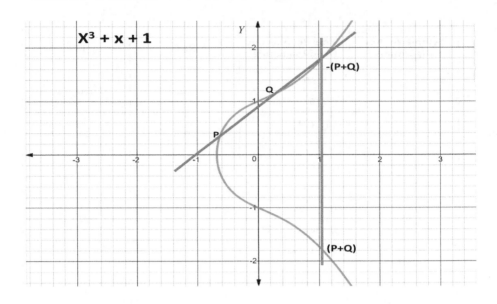

Figure 2.12: Elliptic curve : $x^3 + x + 1$.

Figure 2.12 shows the example of an elliptic curve. The elliptic curve ∇ shows slope of the line L that joins two points $P(x_P, y_P)$ and $Q(x_Q, y_Q)$. Where $\nabla = (y_Q - y_P)/(x_Q - x_P)$. **Point addition** operation in elliptic curve can be represented as a $R(x_R, y_R) = P(x_P, y_P) + Q(x_Q, y_Q)$ and **point doubling** operation on the elliptic curve can be defined as a $R = P+P = 2P$. Point addition is calculated in normal algebra as follows. If $P(x_P, y_P)$ then negative of P is $P(x_P, -y_P)$ and which can be notified as -P. $P + -P = O$ (point of infinity).

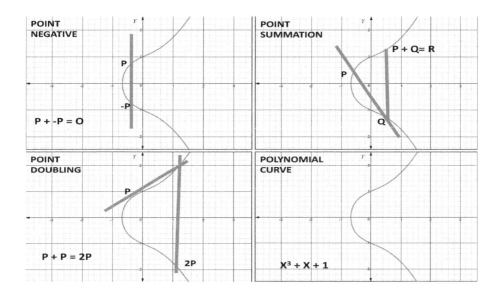

Figure 2.13: Elliptic curve operation.

Figure 2.13 shows how the elliptic curve operations are going to perform.

$$x_R = \nabla^2 - x_P - x_Q \quad y_R = -y_P + \nabla(x_P - x_R) \tag{2.3}$$

and point doubling calculated in normal algebra is as follows.

$$x_R = (\{3x_p^2 + a\}/2y_P) - 2x_P \quad y_R = (\{3x_p^2 + a\}/2y_P) * (x_P - x_R) - y_p \tag{2.4}$$

2.4.1 Elliptic curve over prime field

With elliptic curve over prime field, we consider all the values of variable and co-efficient in the set of integers ranging from 0 to p-1. Extra operation in the prime field is mod p operation. A cubic equation for the prime field is:

$$y^2 \bmod p = (x^3 + ax + b) \bmod p$$

For example, if a = 1,b = 1, x = 1, y = 7, p = 23, then we can say that:

$$7^2 \bmod 23 = (1^2 + 1 + 1) \bmod 23$$

All the values of a and b must satisfy following equation in the prime field.

$$(4a^3 + 27b^2) \bmod p \neq 0 \bmod p$$

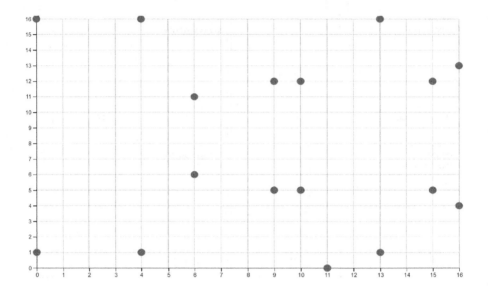

Figure 2.14: Elliptic curve : $x^3 + x + 1$ **mod 17.**

Figure 2.14 shows points on the curve $x^3 + x + 1$ mod 17 where a=1,b=1. Points other than point of infinity O are (0,1), (0,16), (4,1), (4,16), (6,6), (6,11), (9,5), (9,12), (10,5), (10,12), (11,0), (13,1), (13,16), (15,5), (15,12), (16,4), (16,13). Number of points here are equal to number of points in Z_{17}. Negative of any point P is denoted by -P; for example, in $E_{17}(1,1)$ for P=(9,5) a negative is -P(9,-5). But -5 mod 17 is 12 and $(9,12) \in E_{17}(1,1)$ mod 17. Point addition is defined in prime field as follows. Let us take $P(x_P, y_P)$ and $Q(x_Q, y_Q)$. Now if $P \neq Q$ then:

$$\nabla = ((y_Q - y_P)/(x_Q - x_P)) \bmod p$$

and if P=Q,

$$\nabla = ((3x_P^2) + a/(2y_P)) \bmod p$$

Now let us take $R(x_R, y_R)$ = P+Q. then:

$$x_R = (\nabla^2 - x_P - x_Q) \bmod p$$
$$y_R = (\nabla(x_P - x_R) - y_P) \bmod p.$$

Multiplication is nothing but repeated additions; for example, 3P = P+P+P.

2.4.2 *Elliptic curve over Extension field*

Cubic operation for the extension field $\mathbb{GF}(\nvDash^>)$ then other elliptic curve fields. Elliptic curve over extension field is also considered an elliptic curve over binary field.

Cubic equation for extension field is :

$$y^2 + xy = x^3 + ax^2 + b \tag{2.5}$$

Extension field elliptic curve can be signified as $E_{2^m}(a,b)$ and variables x, y and coefficient a and b are elements of $E_{2^m}(a,b)$. All the operation in the $E_{2^m}(a,b)$ will be mod 2 operations.

Let $P(x_P, y_P)$ and $Q(x_Q, y_Q)$ are two points on $E_{2^m}(a,b)$. A negative of point P in the binary field is -P but it is different in terms of calculation. If $P(x_P, y_P)$ then $-P(x_P, x_P + y_P)$ is the negative of P. Now if $P \neq Q$ and -Q means point summation, then $R(x_R, y_R)$ is calculated as follow:

$$\nabla = (y_Q + y_P)/(x_Q + x_P)$$
$$x_R = \nabla^2 + \nabla + x_P + x_Q + a$$
$$x_R = \nabla(x_P + x_R) + x_R + y_P$$

Now let us consider point doubling operation. So R = 2P is calculated as follow.

$$\nabla = x_P + (y_P + x_P)$$
$$x_R = \nabla^2 + \nabla + a$$
$$y_R = x_P^2 + (\nabla + 1) * x_R$$

2.4.3 Elliptic curve Diffie-Hellman problem(ECDHP) and Discrete logarithm problem (ECDLP)

Elliptic curve Diffie-Hellman problem [Menezes et al. (1996)] can be seen as a normal Diffie-Hellman key exchange problem with different public and private elements. Let elliptic curve $E_p(a,b)$ be defined in which p is prime number greater than 3 or the number defined as a 2^m. Now let $P(x_P, y_P)$ is the generator point.

Then in the ECDHP, Elliptic curve discrete logarithm problem is defined based on given K_{PU_B} and $P(x_P, y_P)$, computing $K_{PR_B} = P(x_P, y_P) * K_{PU_B}^{-1}$ is computationally unfeasible, which means it is a very hard thing to calculate a private key in polynomial time.

- $E_p(a,b)$ means curve equation and a generator point P is public elements.

- Alice chooses its private key $K_{PR_A} \in \{0,1,2,.....p-1\}$

- Bob chooses its private key $K_{PR_B} \in \{0,1,2,.....p-1\}$

- Alice computes $K_{PU_A} = K_{PR_A} * P(x_P, y_P)$ mod p(Discrete logarithm problem)

- Bob computes $K_{PU_B} = K_{PR_B} * P(x_P, y_P)$ mod p(Discrete logarithm problem)

- Both Alice and Bob shares K_{PU_B}, K_{PU_A} with each other

- Alice computes $K_{A,B} = K_{PU_B} * K_{PR_A}$

- Bob computes $K_{A,B} = K_{PU_A} * K_{PR_B}$

- Value of $K_{A,B}$ computed by Alice and Bob will be similar.

2.4.4   Message Encoding and Decoding as Elliptic curve points

Alice wants to share message M with Bob, and curve defined is $E_p(1,1)$ then let us say Alice wants to generate a M in to form of point $P(x,y)$ on the curve $x^3 + x + 1$.

- Alice chooses plain text and converts it into decimal number by alphabetic theory.

- Alice chooses one public variable k and computes $x = m * k + i$ where value of i will be ranging from $1 top - 1$

- Alice takes i = 1 and computes x, if value of x satisfy $y^2 = x^3 + x + 1$ mod p and value of y as a quadratic residueperfect square root. Then consider pair (x,y) else increment the value of i and repeat until we get a valid pair (x,y). This pair (x,y) is point encoding of message M.

- To decode the M from a given point (x,y), consider following equation:

m = floor((x-1)/k)

2.4.5 Message Encryption and Decryption on Elliptic curve points

Message encryption in elliptic curve cryptography is an important part of communication. For the encryption and decryption algorithm, we assume that both the communication parties have securely transmitted keys and generated $K_{A,B}$. Now let Alice wants to send message M to Bob, so Alice generates a cipher text of M as follow:

■ Alice chooses random integer k and computes cipher text $C_m = \{k * P, M + k * K_{PU_B}\}$ and transmits to Bob.

■ Bob receives C_m and extracts $k * P$ and $M + k * K_{PU_B}$.

■ Now Bob takes $M + k * K_{PU_B} - k * p * (K_{PR_B}) = M + k * (P * K_{PR_B}) - k * p * (K_{PR_B}) = M + (k * p * K_{PR_B} - k * p * K_{PR_B}) = $ M.

Elliptic curve cryptography is the most suitable cryptography protocol for the emerging authentication techniques for devices where power backup is low and computation capabilities are less.

2.5 Scalar multiplication problem

The scalar multiplication problem defines implementation of elliptic curve cryptography in hardware and software. Elliptic curve cryptography involves a number of mathematical operations in implementation. It involves summation, multiplication, square and inverse operations. Total number of operations defines the complexity of the problem. Most significant bit(MSB) is the left most bit in the binary representation, and the least significant bit(LSB) is the right most bit in the binary representation. Let $P(x_P, y_P,)$ represents the two points on the elliptic curve $y^2 + xy = x^3 + ax + b$ defined over $\mathbb{GF}$ $\not\Vdash$ $^\triangleright$. We have already seen its operation elliptic curve section. Now we want to compute $R = k * P$ where k is any integer number and P is a point on the elliptic curve. We can compute value of R using scalar multiplication using MSB first bit, LSB first bit and Montgomery scalar multiplication operation.

2.5.1 Scalar multiplication MSB First bit

Let us take point $P(x_P, y_P)$ defined over elliptic curve and we compute $R = k * P$.

■ Step 1: Convert k in to binary form $(k_{m-1}, k_{m-2}, \ldots, 0)_2$. where k_m is the most significant bit with value 1

Algorithm:

■ 1. Q = P

■ 2. for i = m-2 to 0

■ 3. Q=2*Q

■ 4. if k_i = 1 then

■ 5. Q=Q+P

■ 6. end if

■ 7. end for

■ 8. return Q

Example: Let P = 7 and k = 7 and we want to compute $k*P$.

■ Step 1: k = $(111)_2$

■ 1. Q = 7

■ 2. i = m-2, Q = 14, k_{m-2} = 1, Q = Q+P = 21.

■ 3. i = 0, Q= 42, k_0=1, Q= 42+7 = 49

■ Return Q = 49

Scalar multiplication MSB first algorithm requires m point doubling and $(m-1)/2$ point addition. $7P = 2(2P+P)+P$. Basic logic behind scalar multiplication is first double and after add. It uses only one register for accumulation and doubling operation.

2.5.2 Scalar multiplication LSB First bit

Let us take point $P(x_P, y_P)$ defined over elliptic curve. and we compute $R = k*P$.

■ Step 1: Convert k in to binary form $(k_{m-1}, k_{m-2}, \ldots, 0)_2$. Where k_m is the most significant bit with value 1
Algorithm:

■ 1. Q = 0,R=P

■ 2. for 0 = m-1

■ 3. if $k_i = 1$ then

■ 4. Q=Q+R

■ 5. end if

■ 6. $R = 2 * R$

■ 7. end for

■ 8. return Q

Example: Let P = 7 and k = 7 and we want to compute k*P.

■ Step 1: k = $(111)_2$

■ 1. Q = 0, R = 7

■ 2. i = 0, k = 1,Q = 7, R=14.

■ 2. i = m-2, k =1 ,Q = 21, R=28.

■ 2. i = m-1, k = 1,Q = 49, R=56.

■ Return Q = 49

Scalar multiplication LSB first algorithm requires m/2 point doubling and m/2 point addition. This approach needs two different registers for accumulation and doubling approach. This approach will be most preferable approach when parallel computing is involved in the implementation.

2.5.3 Montgomery Scalar Multiplication Operation

In the Weierstrass point addition, let $P(x_P, y_P)$ and $-P(x_P, x_P + y_P)$ and equation is $y^2 + xy = x^3 + ax^2 + b$, (x,y)$\in 2^m$. Point addition requires one inversion and two multiplication operations. Montgomery noticed that value of y_R in the computation of $R(x_R, y_R)$ is not dependent on the computation of x_R. So he gave an algorithm to perform the scalar multiplication operation [Menezes et al. (1996)].

> ■ Goal is to compute $Z = k * R$. Input $k > 0, R = P + Q$. Where $p = (x_P, y_P)$ and $Q = -P = (x_P, x_P + y_P)$ **Algorithm:**
>
> ■ 1. Set k = $(k_{m-1}, k_{m-2}, \ldots, k_0)_2$.
>
> ■ 2. Set Q = R and P=2P
>
> ■ 3. for i from m-2 to 0
>
> ■ 4. if $k_i = 1$
>
> ■ 5. Set Q = Q+P and P=2P
>
> ■ 6. else
>
> ■ 7. Set P=P+Q, Q=$2 * Q$
>
> ■ Return Z = Q.

Algorithm proposed by Montgomery requires $2(m-2) + 1$ inversions, $2(m-2)/4$ multiplications, $4 * (m-2) + 6$ additions, and $2(m-2) + 2$ square root operations.

2.6 Hash-based operations

Hash calculation is one of most important part of complete cryptography. Hash is the fixed size finger print or digest of message that can be used to ensure the integrity, authentication, digital signature, random number generation and so on. Lamport in (Lamport (1981)) had for the first time suggested the concept of hash calculation. There are three most important parts of hash-based cryptography. First one is the hash function, which is actually a combination of hash function and size reduction function, hash chain and its calculation and biometric-based hashing.

2.6.1 Hash function

One way cryptographic hash function $H : \{0,1\}^* \to 0, 1^n$. It takes a binary string $p \in \{0,1\}^n$ of any arbitrary length as an input and produces a fixed size n binary string $Q \in \{0,1\}^n$ as output. So hash functions are the auxiliary functions that are used for digital signature, key exchange, random number generation authentication and so on.

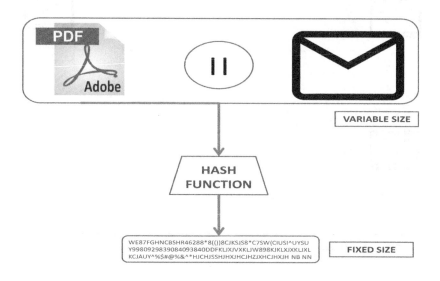

Figure 2.15: Hash calculation and comparison.

In Figure, 2.15 hash generation of email message with pdf file attachment is shown. The following algorithm shows simple digital signature using hash function for mail example.

- Bob generates K_{PU_B} and shares with Alice

- Bob writes a mail message(x) and applies the hash function so $h_B = H(x)$ and generates the digital signature using his private key, $DS_B = sig_{K_{PR_B}}(h_B)$.

- Alice receives message x and digital signature DS_B, Alice computes $h_A = H(x)$ and verifies the signature using Bob's public key, applies verification function, $verify_{K_{PU_B}}(h_A, DS_B)$.

Hash function itself does not assure the confidentiality of message. So hash function used with many encryption algorithms ensure the confidentiality, integrity, authentication and other parameters[Stallings (2010)].

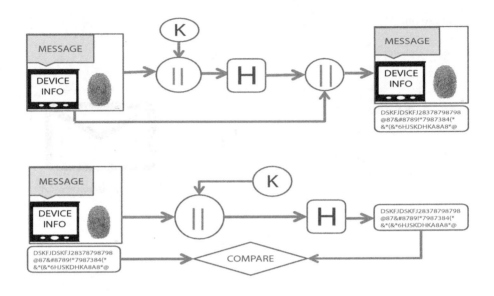

Figure 2.16: Hash calculation and comparison.

Figure 2.16 shows the calculation of hash function using pre-shared key between sender and receiver. It is one way symmetric hash function, in which hash function calculates hash using key K and later sender sends message concatenated with hash output. Receiver receives plain text message and hash function. So receiver also calculates hash function using pre-shared key K and compares the generated output with the sender's hash. So this is a simple hash function that works as unsecured check-sum. Here, there are many different open playgrounds for the attackers. An adversary(an attacker) can alter the message and recompute the same size hash and send to the user.

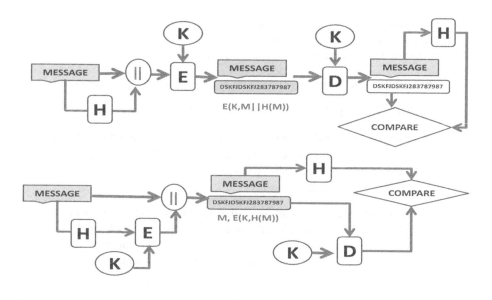

Figure 2.17: Hash-based authentication.

In Figure 2.17, the user computes a hash function and applies encryption algorithms. In the first case, the user encrypts both message and hash function, which ensures confidentiality and integrity of message and hash function. In the second case, the user only encrypts hash function, which ensures integrity but does not ensure confidentiality of message.

A stronger hash function is considered if the single bit change in the message ensures complete hash value. So it is in-feasible to find out a message or data that generates the same hash value for input data or message.

Major applications of cryptographic hash functions include **Authentication** in which message authentication can help to maintain the integrity of the message. Device authentication and user authentication help to authenticate the identity of device and user. In the world of internet of things, billions of devices are involved. So identification and validation of device will be a complex part of security. Hash functions that generate the hash value with the help of some key, seed or nonce is called a keyed hash function or message authentication code function. Other applications of hash functions include one-way password file generation, intrusion detection and virus detection and developing a pseudo random function. A very simple hash function is applying XOR operation for the n times for n bit size input.

Hash function computation involves two important aspects. One is collision and other is preimage. Collision occurs when there are two different messages x and x' but their hash computation is $H(X) = H'(x)$. So its a collision of message, and the second aspect is preimage. Preimage is nothing but an input in the hash function. If p is the number of bits then 2^p total messages are possible and q is the number of bits in hash size then 2^q total hashes are possible.

Any good hash function will satisfy following property:

- **Variable size input:** Any size input will be allowed in hash function.

- **Fixed size output:** Output size will be fixed for hash function.

- **Preimage resistant or one-way property:** If x is the input and hash value $h = H(x)$, then it is computationally unfeasible to find $h = h'$ where $h' = H(y)$. So it is easy to calculate hash from given preimage, but it is very hard to find pre-image from the given hash.

- **Second preimage resistant or weak collision property:** For any give x if $h = H(X)$ then it is computationally unfeasible to find $y \neq x$ where $H(y) = H(x)$. So it is given message x, but it is hard to find the alternate message such that hash value of given message and hash value of an alternate message become same.

- **Collision resistant or strong collision property:** Computationally in-feasible to find a pair of two values x and x' such that value of $H(x) = H(x')$. This property helps to maintain mutual authentication.

- **Efficiency and randomness:** So computation of hash function must be easy, and it should involve randomness in the hash text.

If L is the large message size, then hash function divides L in to b bits size of n blocks. Each block includes length of bit at last to add toughness for the attacker to find same size hash. So intruder needs to find two equal length messages that generate equal length hash. So length helps to satisfy collision resistant property.

2.6.2   Hash chains and its calculation

Lamport in 1981 [Lamport (1981)] introduced concept of hash chain. Hash chain is the use of sequences of the hash functions to improve the security. Figure 2.18 shows the basic hash chain operation. If n times hash function is applied then hash chain can be denoted as $n^{1000}(preimage)$.

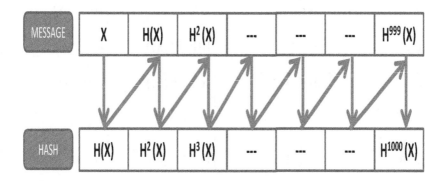

Figure 2.18: Hash chain.

Let us understand computation of hash chain using simple analogy of client server.

■ Client first time registers user name(U) and password(P) with the server.

■ Server decides value of n and computes $h^n(P)$ and stores it in a file. Server does not compute all values, but it stores only the last value.

■ Now when ever user want to log in, it enters user name so server requests to enter i^{th} password, where value of i = 1 for the first time. So user computes hash operation h^{1000-i} times on password and sends to server. Now server computes $h(h^{1000-i})$ on received password from user and stores h^{1000-i}. If $h^{987}(x) = h^{123}(x)$ then given $h' = h^{123}(x)$, one can easily compute $x' = h^{986}(x)$.

PASSWORD SEQUENCE	CLIENT SIDE COMPUTATION	SERVER SIDE COMPUTATION	SERVER STORED VAULE
0	NA	NA	$h^{1000}(x)$
1	$h^{999}(x)$	$h(h^{999}(x))$	$h^{999}(x)$
2	$h^{998}(x)$	$h(h^{998}(x))$	$h^{998}(x)$
....	....	....	....
998	$h^2(x)$	$h(h^2(x))$	$h^2(x)$
999	$h^1(x)$	$h(h^1(x))$	$h(x)$

Figure 2.19: Hash chain computation.

In the above Figure 2.19 computation, even though intruder gets the hash value, he can not use it for the next communication because server computation does not match to stored value. So hash chain is the set of values $h_0, h_1, h_2, h_3, h_4, \ldots, h_n$ for $n \in \mathbb{Z}$, such that $h_i = H(h_{i-1})$ for some hash function H, where $i \in [1, n], x_0$ is valid input for H. So hash chain n is the total number of times hash operation is performed. Hash chain must be easy to compute. Any value x_i can be calculated from any value x_j if value of $i \geq j$. So computation of hash chain makes it suitable tool for platforms with less memory and less computing. If the synchronous between user and server breaks, then user is sending h_j, and system using h_k to authenticate, $j \neq k$ then this can be detected by repeatedly applying hash function to both the values until match is obtained. So if $j > k$ then computes and if $j < k$, requests later values. So if j = 5 and k = 3 than $h_5 = H^2(h_3)$.

2.6.3 Bio hashing

There are famous hash function algorithms prepared by researchers. SHA series and MD5 series are most commonly used algorithms in cryptography. All of these algorithms are useful when the same password is input to the computation every time, but let us take the example of bio-metric based registration where the same user enters more than 10 to 12 different types of input associated with one hash value. So it becomes difficulty to compare. Bio hashing is the special category of hash functions that are invariant to these changes. Bio hashing function first generates user specific random vectors and generates bio code and then represents project coefficient into zero or one.

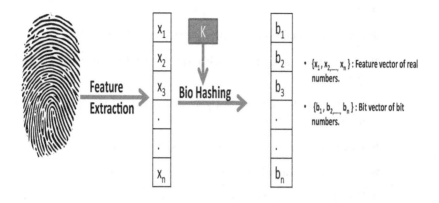

Figure 2.20: Bio Hash.

Calculated efficient is compressed to $b \in \{0,1\}^m$ where m is the length of bit string. Figure 2.20 shows that the image of the hash value will be calculated using feature extraction method. From the input image feature, vector will be generated by processing with the help of wavelet transform, long polar transform, high pass filtering. Bio hashing use in the multi factor authentication scheme. Where multiple parameters like password and biometric or ID and biometric exist. Bio hashing application is critical to implement in the IoT due to complexity increment in the computation. So it may be that researcher need to focus on light weight authentication schemes.

2.7 RSA Algorithms

Ron Rivest, Adi Shamir, and Len Adlesman at MIT gave an algorithm [Stallings (2010)], also called as a RSA algorithm that can be used for encryption/decryption, digital signature, and key exchange. Security of RSA algorithm lies in the prime number factoring. Let p and q be two large prime number; then if we compute $n = pq$ and make n public then it should be computationally unfeasible to find out p and q. A typical plain text block size in RSA is 1024 bits or 309 decimal.

Algorithm can be stated as follow:
In algorithm Alice will first generate a pair of keys that are used by Bob for an en-

cryption purpose.

■ Alice chooses two private large prime numbers p and q.

■ Alice computes $n = p * q$.

■ Calculate Euler totient function $\phi(n) = (p-1)*(q-1)$.

■ Alice chooses integer e in such a way that $GCD(\phi(n), e) = 1, 1 < e < \phi(n)$.

■ Alice computes d in such a way that e*d≡ 1 (mod $\phi(n)$). so d is multiplicative inverse of e.

■ Alice creates pair of public key of his. $K_{PU_A} = \{e, n\}$

■ Alice creates pair of private key of his. $K_{PR_A} = \{d, n\}$

■ Bob has plain text M of size less than n.

■ Bob calculates cipher text $C = M^e$ mod n.

■ Alice has cipher text C.

■ Alice calculates plain text $M = C^d$ mod n.

RSA algorithm works because e and d are multiplicative inverses of each other on modulo $\phi(n)$. It is possible to find e, d, and n so that $M^{e,d}$ mod n = M and this can be only possible if $e * d \equiv 1$ mod $(\phi(n))$. So $d = e^{-1}$ mod $(\phi(n))$. Example of RSA[Stallings (2010)].

Let is take prime number p = 17 and q=11 so n=187 and $\phi(n) = 160$. Now select the e in such a way that $GCD(e, \phi(n)) \equiv 1$. So let us take $e = 7$. Now calculate the value of d in such a way that $e * d \equiv 1$ (mod $\phi(n)$). So we need to solve $7 * d \equiv 1$ (mod $\phi(n)$). We can use the Extended Euclidean algorithm to find the value of d.

■ $7 * d \equiv 1$ (mod 160) = GCD(7,160) = $7 * x + 160 * y \equiv 1$ mod 160 so the value of x will be the multiplicative inverse of 7 (Refer Extended Euclidean algorithm).

■ 160 = 7*22 + 6 ⇒ 6 = 1*160 + (-7)*22.

■ 7 = 1*6 + 1 ⇒ 1 = 1*7 + (-1*6) replace value of 6.

■ 1 = 1*7 + (-1*(1*160 + (-7)*22)) = 23*7 + (-1)*160.

■ So the value of x is 23 and value of y is -1. So value of d and multiplicative inverse of e is 23 which satisfy $e * d \equiv 1$ (mod $\phi(n)$).

■ We can verify correctness of answer by computing 23*7 mod 160= 161 mod 160 = 1.

so the value of d = 23. So the pair of public key $\{e,n\} = \{7,187\}$ which will be used for encryption("e" indicates encryption) and pair used for decryption $\{d,n\} = \{23,187\}$.

Solving the factoring problem is the real toughness of an RSA algorithm. There are so many versions of RSA based on size of n. Attacks like brute force attack, mathematical attack, timing attack, hardware fault based attack, choose cipher text attack.

2.8   Bilinear pairing system

We have seen an elliptic curve based cryptography in previous sections. Bilinear pairing is also an extension of the work of elliptic curve based cryptography. Elliptic curve P256 is the most commonly used curve where P indicates prime finite field and 256 indicates number of bit generator point size. Top websites that use ECDHE work on P256 curve. Now pairing is used for the mapping of two groups in to one group. Pairing is map $e : G_p * G_q \rightarrow G_T$, where G_p, G_q and G_T are the cyclic groups. If the G_p and G_q are similar then it is called symmetric pairing and it G_p and G_q are not same than it is asymmetric pairing. G_T is a target group [Menezes et al. (1996)].

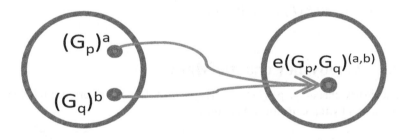

Figure 2.21: Pairing operation.

So as shown in Figure 2.21, It will take few points from source group and match them to target group in such a way that exponents will multiply. Exponents can be single variables, can be a vectors or can be matrix.

Properties of pairing:

- $e(G_p^a, G_q^b) = e(G_p^b, G_q^a)$

- $e(\{G^{a_1}, G^{a_2}.....G^{a_n}\}, \{H^{b_1}, H^{b_2}.....H^{b_n}\}) \Rightarrow e(\{G, H\}^{a,b})$.

- Bi-linearity

- Non degeneracy : If P is the generator of Group G, then $e(P, P)^{(x,y)} \neq 1$.

- Computability : Pairing computation should be easily computable in polynomial time.

- If i is the identity of the group G and $P, Q \in G$ then $e(P, I) = e(I, Q) = 1$.

- $e(P, -Q) = e(Q, -P) = e(P, Q)^{-1}$.

- $e(aP, bQ) = e(P, Q)^{(a,b)}$ for all $a, b \in \mathbb{Z}$.

- $e(P, Q) = e(Q, P)$.

- if $e(P, Q) = 1$ then either P=1 or Q=1.

2.8.1 Bilinear pairing system decision Diffie-Hellman problem

If $G, G^x, G^y, G^z \in G$, then pairing $e(g, g^z) = e(g^x, g^y)$ can help us to identify whether $Z = x * y$ it is decision Diffie-Hellman problem. So given generator P, Alice generates aP, Bob generates bP, and Carol generates cP. So computing $e(P, P)^{a,b,c} = e(aP, bP)^c =, e(bP, cP)^a = e(aP, cP)^b$.

2.9 Chebyshev chaotic Map:

Chebyshev chaotic map $T_n(x) = [-1, 1] \rightarrow [-1, 1]$ is the recurrence relation used in cryptography for authentication purpose.

$T_n(x) = 1$ if n = 0
$T_n(x) = x$ if n = 1
$T_n(x) = 2_x T_{n-1}(x) - T_{n-2}(x)$ if n $\geq$ 2 mod p

Where n is an integer and x is a variable taking values over interval $(-1, 1)$. So Chebyshev polynomial T_n of degree n is defined as

$T_n(x) = cos(n * arccos(x))$ where $x \in [-1, 1]$
$cos(n\theta)$ if $x = cos\theta, \theta \in [0, \Pi]$

Examples of Chebyshev polynomials are:
$T(x) = 2x^2 - 1$

$T(x) = 8x^4 - 8x^2 + 1$

Basic properties of Chebyshev polynomials are semi-group property and chaotic map property [Menezes et al. (1996)].

- **Semi-group property:** Semi-group property of Chebyshev map holds under interval $(-\inf, \inf)$ and defined as $T_n(x) = 2xT_{n-1}(x) - T_{n-2}(x)$ mod p where $n \geq 2$,p is large enough prime number and $x \in (-\inf, \inf)$, $T_a(T_b(X)) = T_b(T_a(X)) = T_{a,b}(x)$ mod p, where a and b are positive numbers,x $\in$ [-1,1] and $p \in \mathbb{Z}_i^*$. Semi-group property of the Chebyshev map helps in many different authentication and key exchange algorithm designing.

- **Discrete Logarithm problem:** Chaotic map based discrete logarithm problem can be defined as, for any given x and y, it is computationally in-feasible (not in polynomial time) to find integer p such that $T_p(x) = y$.

Key exchange and message encryption using Chebyshev polynomial:

Alice Computes:

- Generate a large integer A.

- Select number $x \in [-1, 1]$.

- Set public key $(x, T_A(x))$.

- Set private key A.

Bob computes:

- Receive Alice public key.

- Generate a large integer B.

- Encode message M in the interval of $[-1, 1]$.

- Compute $T_B(x)$.

- Compute $T_{A,B}(x) = T_B(T_A(x))$.

- Compute $Z = M * T_{A,B}(x)$

- Generate cipher text $C = (T_B, Z)$ and send to Alice

Alice recovers message M:

- Compute $T_{A,B}(x) = T_A(T_B(x))$.

- Compute $M = Z/T_{A,B}(x)$.

2.10   Introduction to block chain technology

After the evolution of bitcoin in 2008 [Nakamoto (2008)], block chain technology attracted many researchers. Bitcoin is the cryptocurrency that uses block chain technology as a back end processing. Block chain is the publicly distributed storage that contains encrypted ledger, and block chain technology can be used in many different applications like banking, payment and transfer, health care, law enforcement, real estate and online entertainment. Block chain technology is also used in distributed cloud storage and distributed data processing. Block chain contains link of blocks as a ledger and each block contains some cryptographic parameters and transaction details. Every transaction is in block chain technology broadcasted with every node in the chain. Every node that receives the transaction details do the verification of transaction and identity by signature inside transaction. Signature in block chain technology is the combination of public key and private key of sender. Multiple nodes involved in the approvals of single transactions makes it robust, but at the same time it increases the delay in processing. Block chain technology has three major parts:

1. Public key based cryptography algorithms

2. Peer to peer network

3. Block chain programs

Block chain uses public key cryptography for identity security and hash function to maintain the integrity. Most famous feature of block chain technology is the decentralized approach. There is no central control system, which ensures robustness and scalability. Major challenges involved in the adoption of block chain technology are:

■ Resource constrained devices in the IoT makes computationally intensive mining a major challenge

■ Large number of nodes increases, scalability issues

■ Unknown mathematical computations and protocol increases computation overhead.

Block chain technology involves three major types of block chain:

■ **Public block chain:** Everyone in the internet can see,verify and edit the ledger. Everyone can add a block of transaction in the chain.

■ **Private block chain:** Everyone in the internet can see the transaction details, but limited people can verify and add a block of transaction.

■ **Consortium:**It is a combination of public and private. Only a group of particular organizations can verify and add the transactions.

2.11 Summary

In this chapter, we have discussed basic mathematical foundations required for the designing of cryptography schemes. This chapter includes mathematics behind elliptic curve cryptography and other security algorithms. Starting from basics of linear algebra to advanced elliptic curve computation is discussed in detail. This chapter enables basic knowledge of other cryptographic maths like bi-linear pairing and

Chebyshev chaotic map. It will work as complete mathematics package for the authentication scheme designers.

2.12 References

Hou, G. and Z. Wang (2017). A Robust and Efficient Remote Authentication Scheme from Elliptic Curve Cryptosystem. *19*(6), 904–911.

Lamport, L. (1981, November). Password authentication with insecure communication. *Commun. ACM 24*(11), 770–772.

Luo, M., Y. Zhang, M. Khan, and D. He (2017). A secure and efficient identity-based mutual authentication scheme with smart card using elliptic curve cryptography. *International Journal of Communication Systems* (February).

Menezes, A. J., S. A. Vanstone, and P. C. V. Oorschot (1996). *Handbook of Applied Cryptography* (1st ed.). Boca Raton, FL, USA: CRC Press, Inc.

Nakamoto, S. (2008). Bitcoin : A Peer-to-Peer Electronic Cash System. pp. 1–9.

Stallings, W. (2010). *Cryptography and Network Security: Principles and Practice* (5th ed.). Upper Saddle River, NJ, USA: Prentice Hall Press.

Zhang, L., S. Tang, J. Chen, and S. Zhu (2015). Two-Factor Remote Authentication Protocol with User Anonymity Based on Elliptic Curve Cryptography. *Wireless Personal Communications 81*(1), 53–75.

Chapter 3

IoT Authentication

CONTENTS

√ In god we trust, All others must bring data.

W. Edwards Deming
Statician

3.1   Abstract

Authentication in cryptography is an important parameter in terms of foolproof security. Authentication enables communicating parties to verify their identity and their correctness. In the internet of things, authentication between user - user and device - user will play important role to ensure complete secrecy. So in this chapter we have discussed IoT authentication and various scenarios in IoT where we will need to implement authentication protocols.

3.2   Authentication layered architecture

In the Chapter 2, we have highlighted basic mathematics behind cryptography and its methods, which can help in the creation of new cryptographic schemes to provide security to the internet. A report produced by inter agency International Cyber Security Standardization Working Group of NIST [Yaga et al. (2018)], titled: "Status of International Cyber Security Standardization for the Internet of Things (IoT)" enlightened various cryptography issues in the internet of things. In Chapter 1, we discussed IoT security lexicon, in which authentication is one of the most critical aspects to achieve. A secure authentication algorithm ensures properties like anonymity, untraceable, privacy, confidentiality, availability and integrity [Aslam et al. (2016)].

In this chapter we discussed introduction to authentication, need of authentication in the internet of things, authentication issues at each layer of IoT architecture, various phases of generalized authentication schemes, and application wise authentication requirements. Number of devices or things in the internet is growing very fast. As per statistica, total number of devices will be 75.44 billions by 2025. This scalability increase, also comes up with various other issues like reliable communication between two devices, Secure communication between two devices and so on.

Reliable communication between two devices is only possible if the communication is protected from active and passive attacks. Common cyber attack possible on IoT devices are physical attacks in which attacker damages physical devices deployed on site, denial of service attacks in which attackers attack either physically or logically to reduce or stop the service, access attack in which attacker tries to gain control of devices [Abomhara and Køien (2015)]. So proper authentication schemes can help to achieve primary goals of security, confidentiality, integrity, availability.

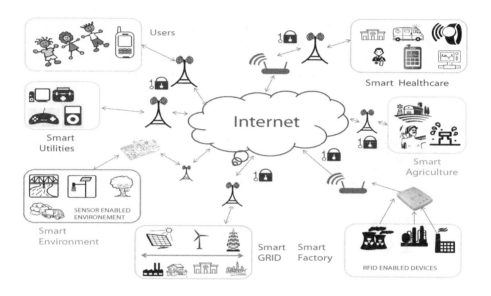

Figure 3.1: IoT Security generalize scenario.

As per highlighted in [Abomhara and Køien (2015)], we can classify IoT devices based on size of device, source of power supply, automation, mobility, IP enabled or not, RFID enabled or not and so on. So as per shown in Figure 3.1, the internet is the center point of services provided by Internet of things. Internet of things comes up with already available older issues in the internet and also with the fresh challenges. A relation between adversary and device capabilities is the evolution in the technology and we can see this in the following way.

The history of computing devices started in 1837, when Charles Babbage proposed first analytical engine. In 1947, Williams filed a patent on CRT based storage system and the journey of RAM started from there. In 1951, Forrester applied for patent on magnetic core memory. In 1969, Intel released first product 3101 schottky TTL bipolar with 64 bit RAM and 1024 bit ROM. And today we can have RAM of more than 16 GB and ROM of 4 TB in our personal laptop, so as shown in Figure 3.2. In the earlier days of computation, sender, receiver and attacker had limited computation capabilities. Various parameters to consider are computation capabilities, RAM, ROM, Clock cycle, Gate Area, Frequency, scheduling algorithms, parallel processing.

In the current scenario we can see that sender machines like laptops, mobiles and receiver machines like servers are capable enough to process complex cryptographic algorithms at the same time attackers also have extended capabilities to perform attacks. Attack vector in the internet is very limited due limited types of devices involved. But attack vector in the internet of things will expanded hugely due to the many different types of devices involved in computations. Most of the internet of things devices will be automated, less power capable, less memory capable. In the internet of things, capabilities of sending device and receiving device will be very limited, but the capabilities of an attacker need not to be limited.

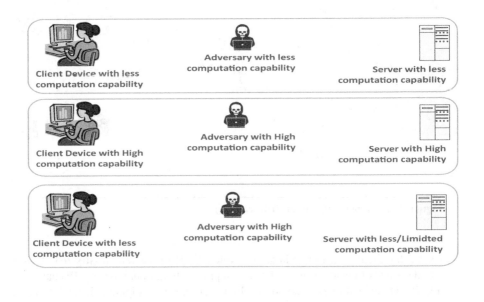

Figure 3.2: Device capability evolution.

So a strong and foolproof light weight authentication and cryptographic algorithm will be required. Recently MIT researchers prepared one hardware chip to perform public key encryption that consumes 1/400 less power consumption, 1/10 memory consumption and is 500 times faster compared to existing cryptography protocol. This chip is prepared based on elliptic curve cryptography. They have used advantage of modular arithmetic operation using 256 bit prime numbers so it reduced memory and power consumption. We will discuss devices, authentication vector and authentication schemes in more detail in the upcoming subsections of this section.

3.2.1 Introduction

Authentication can be defined as a "Method of assuring the identity of the counter part as well as maintaining the integrity of the message and privacy of both communicating entities". There are many authentication schemes proposed so far. The first authentication scheme was proposed by Lamport in 1981 [Jones and Lamport (1981)], based on one time password. In this scheme he used one way hash operation and chaining of hash operation. After that Shimizu in 1991 [Shimizu (1991)] proposed a dynamic password based authentication scheme. Authentication schemes can be divided based on number of factors used, based on mathematical operations used, input parameter based, number of servers involved and so on. In this section, we will go though generalized authentication architecture for the internet of things.

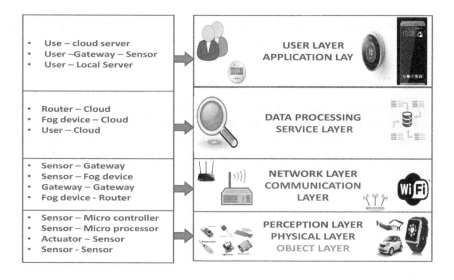

Figure 3.3: IoT Authentication Layered Scenario.

As shown in Figure 3.3, IoT architecture is divided into four different layers,

∎ **Physical Layer:** This layer deals with physical devices that are deployed on the ground and perform data collection activities.

■ **Network Layer:** This layer deals with data aggregation, local data processing, and data forwarding activities.

■ **Data Processing Layer:** This layer deals with cloud computing, data mining and big data processing operation.

■ **User Layer:** This layer deals with mobile applications that read data from cloud or sensors via gateways.

At each layer, different types of devices are involved in communication. such devices are involved in

■ **Physical Layer:** are sensors, actuators, micro controllers, micro processors.

■ **Network Layer:** are gateways, routers, fog devices.

■ **Data Processing Layer:** are cloud storage servers.

■ **User Layer:** mobiles, laptops, computers, actuators.

So authentication scenarios and authentication vectors are different at each layer of IoT. In the subsequent sections we discuss layer wise authentication scenarios.

3.2.2 Physical layer authentication

Physical layer in the internet of things is the backbone of a complete IoT system. More than 60% of IoT devices will be deployed at this layer. And that is why this layer becomes the most critical point for IoT services. Physical layer devices are sensors, actuators and micro controllers or processors that work as a middle point between physical layer and network layer.

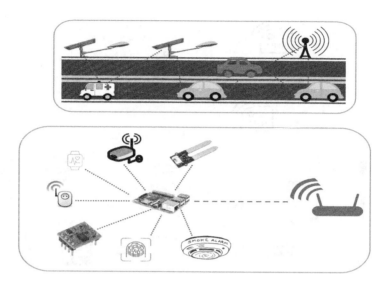

Figure 3.4: Physical Layer Authentication Scenario - 1.

Figure 3.4 shows very generalized scenario of physical layer communication. First part shows scenario of intelligent transport system in which road smart lights, smart vehicles and wifi capable towers are doing communication with each other. Smart lighting continuously senses the data about road conditions, pollution level, fog level, temperature and so on. It changes the light frequency based on the requirements at the same time it also forwards all this data to cars. So cars can also get the indication and behave the someway. Similarly, second part of the Figure 3.4 shows the deeper level of this communication. It shows sensors are communicating with micro controllers and micro controllers are communicating with the network layer gateway.

Major communication protocols in this layer are RFID, NFC, Zigbee, Bluetooth low energy (BLE) and z-wave. IoT device will get the identity based on whether they are IP-enabled or not enabled. IP enabled devices can get the identity of IPv6 and devices that are not IP-enabled communicate using RFID or near field communication (NFC) or Bluetooth low energy (BLE) . As shown in Figure 3.5, RFID tag enabled devices send the information regarding status and availability using 96 bit electronic product code. RFID readers capture this data and forward it to micro-controllers.

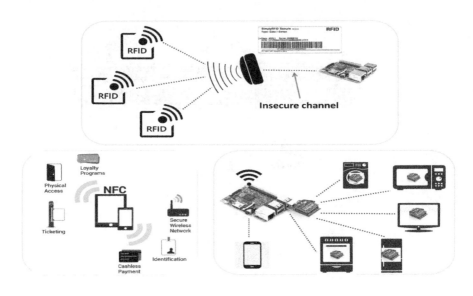

Figure 3.5: Physical Layer Authentication Scenario - 2.

Communication between RFID tag and RFID reader can work by using a secure channel, but most of the communication between RFID reader and micro-controllers or RFID reader and router will be through secure channel. Authentication between RFID reader and gateway becomes difficult due to high scalability. RFID based authentication scheme is discussed in [Wang and Ma (2012)]. In this paper authors discussed untraceable RFID authentication scheme between RFID tag and RFID reader. Another authentication scheme is discussed between RFID reader and transponders in [Bliman et al. (2015)]. Second scenario in Figure 3.5 shows communication between NFC and other NFC enabled devices.

Due to its contact less and faster property NFC is also a good option for communication within short range. One of the secure authentication schemes based on NFC is discussed in [Chen et al. (2010)]. Third scenario in Figure 3.5 shows zigbee communication in which home equipment like TV, washing machines are communicating with micro-processor using Zigbee protocol. Mobile devices connected using IP protocol or NFC or BLE can communicate with micro-processor and capture the direct data. Authentication of valid mobile devices and allowing access of data capturing is a critical aspect. A secure authentication scheme between mobile device and microprocessor is necessary. So authentication at physical layer ensures that ground level devices communicate with each other in a secured manner and does not disclose identity or any other information regarding product or person.

3.2.3   Network layer authentications

Network layer in the internet of things performs combined operation of network layer and transport layer of OSI Model. Network layer ensures reliable and error free communication. There are two protocols in OSI Model; one is IPSec and DTLS ensures security in the network and transport layer. Here in the network layer all WIFI communication protocols, 3G/4G/5G protocols works at this layer. This layer works as middle layer between IoT device and cloud storage. It aggregates the data from various micro-processors or tag readers and forwards it to cloud storage for further processing. Devices at network layers are routers, gateways and fog devices.

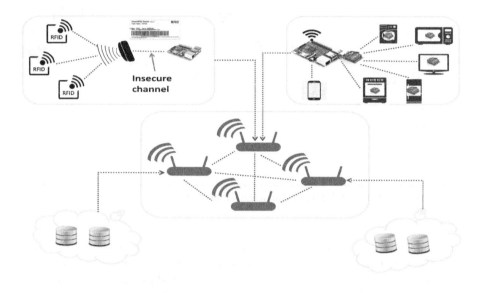

Figure 3.6: Network Layer Authentication Scenario - 1.

Figure 3.6 shows communication of network layer. Microprocessors collect the data and forward the data to the routers. Using various routing algorithms, routers store these data in cloud storage. Here scenarios vary from application to application. Sometimes we need to store the data in local storage and do the processing. Cisco came up with a concept of fog computing. Fog devices are devices those are capable enough to store and do the data processing. With this capability enhancement, fog devices open the grounds for the attacker to capture the data. Authors in [Mukherjee et al. (2017)] discussed major security challenges residing in fog computing. Major characteristics of fog computations are low latency and location awareness, end device mobility, heterogeneity, real time applications, wireless access and capabilities to handle large number of nodes. Authentication is a major problem in fog computing due to large scalability.

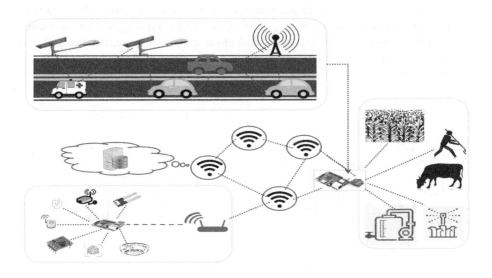

Figure 3.7: Network Layer Authentication Scenario - 2.

Figure 3.7 shows an application oriented scenario of network layer communication. Intelligent transport system routers or the wifi gateway transmits the received messages for processing and storage to the fog devices and routers. Fog devices receive the message and do processing on the message and communicate with other applications as well as store the data in the cloud. Authentication in network layer includes authentication scenarios like authentication between fog devices, authentication between microprocessor and fog device, authentication between micro processor and cloud storage. Some of the application scenario includes user mobile authentication with cloud server via fog devices. So this heterogeneity makes it interesting for researchers to make secure communication protocols or secure hardware that can be attached with fog devices to perform cryptographic operations.

3.2.4 Data processing layer authentication

Data processing layer combines IoT with cloud computing and local data processing.

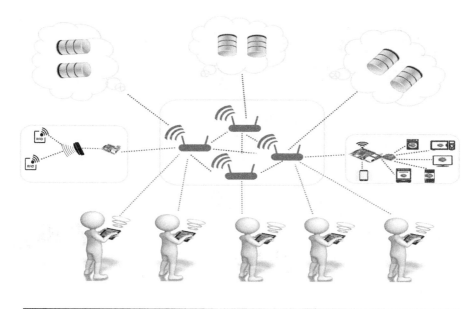

Figure 3.8: Data processing Layer Authentication Scenario.

As shown in Figure 3.8, the data processing layer receives data from the vast number of applications and does the processing on each application data bit. Some of the applications include smart city, which combines other major applications include smart home, intelligent transport system, smart environment, smart grid, smart gas and water distribution, smart industry, smart health and so on. For the applications like smart city, data aggregation and data processing become a critical task. Authentication between users and cloud service becomes more difficult due to heterogeneity, scalability and mobility. Authentication at this layer includes other important parameter called access control, which data of the product of person is allowed to access and by whom. Authentication between user and cloud via gateway and authentication between gateway and cloud is a major challenge in this layer. As mentioned in [Chang and Choi (2011)], major weaknesses in cloud computing are maintaining ID/password, public key infrastructure, multi factor authentication, and single sign on. Cloud computing requires strong access control mechanism and secure authentication mechanism to ensure major security parameters.

3.2.5   *Application layer authentication*

Application layer authentication includes user authentication with applications.

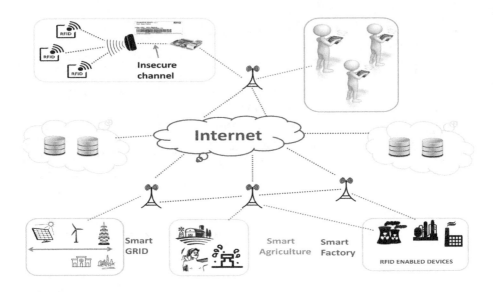

Figure 3.9: Application Layer Authentication Scenario.

As shown in Figure 3.9, multiple users try to capture multiple applications. Every user in the internet of things will have different access on different applications. This heterogeneity makes IoT user authentication more interesting. We can understand application layer authentication by following examples:

■ A user who is owner of home has accessed electricity usage at his home but at the same time user is not authenticated to see details about neighbor's power usage.

■ If the same user wants to get the details about his industry then he must have access to power consumption details of each unit of industry.

■ User in the city is authenticated to see the details about data generated from his home in the smart city, but is not authenticated to see the data of other homes, but as a commissioner of city, user can have access about the data of each home, office and industry.

So this huge heterogeneity creates more complexity at this layer. This layer does the communication with the cloud as well as router as well as micro-processors based on application and requirement of applications.

3.3   IoT node authentication

3.3.1 Introduction

Authentication in the internet of things can be considered as a mechanism to protect:

- Privacy of identity of device, person or any thing.

- Integrity of device related information, person related information and any other data or control signals.

- Access from unauthorized device or persons.

Major authentication schemes or protocols we can divide as follow:

- Password based mutual authentication scheme

- Identity based mutual authentication scheme

- Biometric based mutual authentication scheme

- Key management and key distribution center based centralized authentication scheme

- Light weight authentication scheme

Light weight authentication schemes can be based on

- Elliptic curve cryptography

- Block ciphers

- Stream ciphers

- RFID and bar-code based schemes

- XOR, AND, OR Based authentication schemes

- Or based on any other new methods like block chains.

We can divide IoT nodes in to three major classes [Schmitt, Kothmayr, Hu, and Stiller (Schmitt et al.)]:

- **Type A devices:** Devices like sensors are type A devices, which have RAM of less than 10kByte and Flash of 100kBytes. It is very difficult to implement cryptography mechanism on type A devices due to low capabilities. Most of the devices are battery powered or self powered. It is not possible to provide continuous power supply and add extra memory for cryptographic operation. Type A devices are also called tiny nodes.

- **Type B devices:** Devices like micro-processor or micro-controller are considered class B devices, which have RAM of around 50kBytes to 1 Gb and storage capabilities can be extended up to 32 GB. These devices can be power supplied or battery powered. It depends on its application. We can easily implement some of the light weight authentication algorithms. Major parameters that we can consider for light weight protocols are latency in connection establishment and data transfer, memory consumption, and energy consumption.

■ **Type C devices:** Devices like routers, gateways, fog devices are considered Class C devices that implement various communication protocols like DTLS, IPv6, CoAP, MQTT, XMPP and communicate with user applications. These devices are capable enough to implement complex cryptographic algorithms like DES, AES and complex hash operations like SHA, MD5. The major challenge for these devices is to perform authentication in a high mobility environment. When the user changes home agent to foreign agent and foreign agent to home agent, it is difficult to provide low latency services in a high heterogeneity environment.

3.3.2 Architecture

In this section, we will discuss basic architecture of IoT device authentication scenarios.

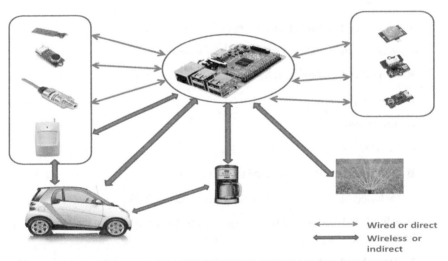

	Wired or direct
	Wireless or indirect

DEVICE TO DEVICE AUTHENTICATION [SHORT RANGE APPROACH]

Figure 3.10: Short Authentication Scenario.

We can view device authentication in two parts based on distance between devices. Either they are directly connected devices or they are connected via some gateways. In major client server architecture, it is not necessary that client and server are directly connected nodes. Client and server authentication may pass through various gateways. Let us understand both the scenarios in very generalized way:

As shown in Figure 3.10, a short range communication environment where direct communication between two authenticating devices is possible. Example: A communication between micro-processor and smart car or communication between smart car and smart office device controller. So this authentication environment will be different than long-range communication. The major variation will be protocols required for communication. For the short range we can communicate via protocols like Zigbee, NFC, BLE, CoAP, MQTT. Cryptographic protocols for authentication can be different due to limited resources at sender and receiver. In this type of environment, for the authentication and key exchange purpose, implementing server or key distribution center is not possible. So mutual authentication schemes can be a suitable option for it. Now let us discuss another environment where client and server distance is far away.

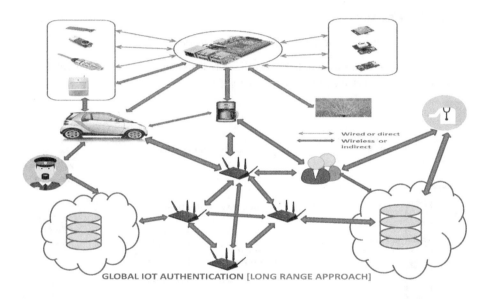

GLOBAL IOT AUTHENTICATION [LONG RANGE APPROACH]

Figure 3.11: Long Authentication Scenario.

Figure 3.11 shows a different environment, where multiple users located at remote areas are trying to communicate with each other. Authentication between doctor and wearable of person driving in some remote area is a real challenge. Communication protocol like DTLS, IPsec, 6LoWPAN, Wifi need to improvise with some additional security features. Another challenge here is a combination of resource constrained device and resource full device. So defining one-way authentication and two-way authentication is a major challenge. To implement any authentication architecture, the following phases need to be implemented:

- ▪ System setup phase also called initializing phase

- Sensors and user registration also called device registration

- Key exchange phase

- Log in and authentication phase

- Password update phase

- Device revocation phase

- Topology updating phase

Let us discuss each phase in a generalized way.

3.3.3   Phases

3.3.3.1   System setup phase by gateway node

This can be the first phase in the generalized IoT authentication environment. Let us understand this phase by the example scenario.

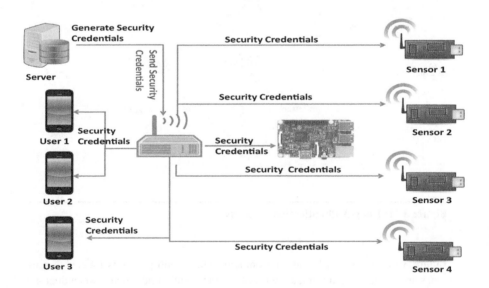

Figure 3.12: Initialization and setup Scenario.

As shown in Figure 3.12, let router or gateway or fog device be connected to micro-processors, four sensors, and server. Server can be the local server or global server. After the topology establishment server shares, then comes the predefined security parameters like random numbers or seed values or prime numbers or any other value. This value will be stored by each connected device and can be used for the input of authentication algorithms. This values will not be available with all the devices, but it will be available to only those devices that are connected during the initialization of topology.

3.3.3.2 Sensors and user registrations

This is one of the complex and important phases in the internet of things due to heterogeneity of the number of devices, flexibility of topology setup and scalability of number of devices.

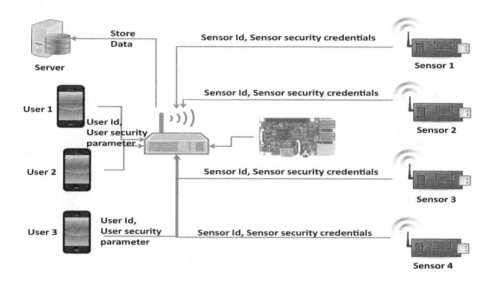

Figure 3.13: Sensor and User registration phase.

As shown in Figure 3.13, every device involved in the communication needs to register with the server if the authentication is via registration server or key distribution center. Every device sends its identity (Ex. RFID Tag number or product id) and some security parameters generated to the server. Majorly authors have proposed authentication scheme based on mutual authentication, single server based authentication and multiserver based authentication. Choice of sin gle server, multi server, or mutual authentication depends on the application and number of devices involved. Another important parameter in IoT environment is dynamic topology and identity. Example: In the manufacturing unit, every product will have a unique identity but the same product will be blocked in box than multiple product will communicate with unique id. So this complexity increases difficulty as well as creates opportunity to think about this type of environment.

3.3.3.3 Key exchange phase

Secure key transfer is one of the most important phases of any authentication schemes. There are good key exchange schemes available like Diffie-Hellman key exchange algorithm or elgamal cryptography. Fewer number of steps and lower number of bit transfers makes any key exchange scheme attractive for the IoT environment.

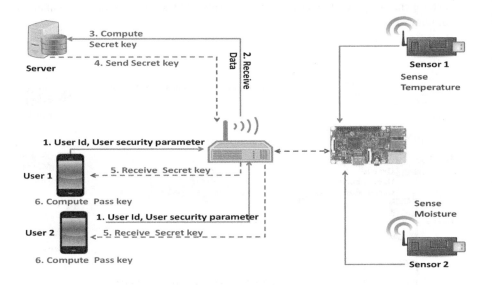

Figure 3.14: Key Exchange phase.

As shown in Figure 3.14, key exchange scheme traverses through the following steps.

1. Devices will send device id and secure parameter(after applying some operation based on scheme) generated to the server; this generated parameter will not be shared with any one, which means it transmitted securely.

2. Server receives a parameter from each device. For devices like the micro-processor (even though it connects multiple sensor), micro-processor will send only its own id and parameter.

3. Based on received parameter, server computes secret key or session key or public key for devices.

4. Server will send this key either securely or make publicly available and store it for authentication.

5. Each device will receive this key.

6. Device computes pass key from received key value or makes use of same key as a pass key.

Above steps discussed are very generalized steps, and it depends on the authors how to make use of it. Some of the key exchange schemes for the IoT environment are discussed in [Li and Liu (2017)].

3.3.3.4 Login phase and authentication

In this phase devices or users make use of device id and generated passkey for to get access to sensors or data.

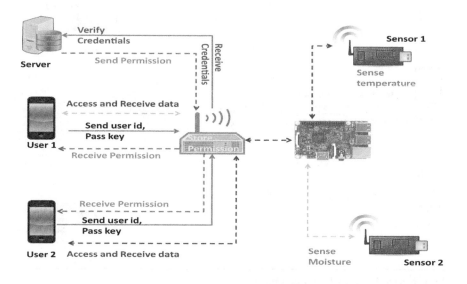

Figure 3.15: Login phase.

Figure 3.15 shows generalized login and authentication phase:

1. User makes use of user id and password to get access from the server or gateway.

2. Server or gateway receives a credential and verifies this credential with stored data base. In the access control mechanism, identity of the device or user plays an important role. Example: If the identity of the family doctor is verified then full access of patient data is available but if the identity is of a specialized doctor, then limited access.

3. Server sends successful message and permission. This permission per id will also be shared with gateways.

4. Device receives permission and requests accessing the data to the gateway.

5. Gateway verifies identity and grants the access.

3.3.3.5 Password update phase

Password update phase is important because the attack chances are higher and for security reasons, it is expected to change password by user or devices.

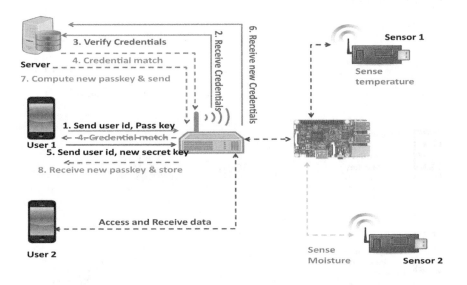

Figure 3.16: Password Update phase.

Password update phase includes the following steps as shown in Figure 3.16:

1. Device will send current id and password to the server.

2. Server will receive the credentials and do the computation.

3. Server will verify the credential; if credentials match then go to step 4, or else reject.

4. If the credentials match,

5. Devices will enter same user id and a new secret key is calculated using received secret key from server.

6. Server will receive this new secret key or security credential.

7. Server will now compute the new passkey and send it to the device in secret environment.

8. Device will receive this new passkey and store.

3.3.3.6 Device adding phase

This phase is important in the dynamic topology environment.

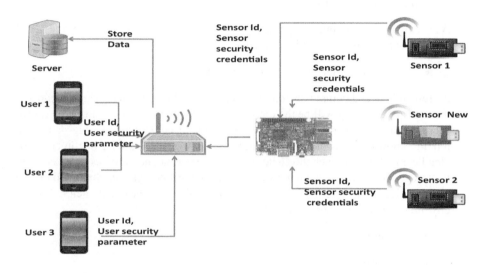

Figure 3.17: New device adding phase.

As shown in Figure 3.17, if new sensor or new device enters in to system, then it has to register with the server via gateway or micro-processor. In the local server environment, implementing this phase is easy compared to a centralized server environment or cloud environment; mobility management is also important aspect in this phase. Authentication between servers or between home agent and foreign agent is also needed. Implementation of this step in a faster and lighter way is a major challenge for IoT environment.

Above all, the phase is used in different ways with different operations and different inputs. But the basic flow of the phase remains the same. We have discussed a scenario where the local server or global server is working as a center point of key exchange and authentication, but in a real time scenario, some larger applications like smart city we would need to authenticate multiple people in multiple applications at different locations. Let us discuss a scenario where the server is working as a center point of communication, also called as a cloud centric approach.

3.4 Cloud centric IoT authentication architecture

Cloud computing [Wang and Ma (2012)] is one of the top ten technologies that will change the world by 2025. Cloud computing works as a centralized data aggregator for many different applications, Major services provided by cloud computing are:

1. Platform as a service

2. Software as a service

3. Infrastructure as a service

As a famous networking company, new web services will be started in the cloud computing environment, and it is called a

■ Sensor as a service or device as a service

As researchers and companies report, if internet of things is successfully deployed, then cloud computing will be the top most service that may have contributed to that success. Various other technologies relate to cloud computing like big data, data mining, knowledge generation for machine learning and artificial intelligence. So cloud computing becomes the center point of operation and cloud as a center point of storage. So providing a strength with advanced technologies in cloud will enable more services in the internet of things. Let us take the example of smart city: A smart city project involves other projects like:

■ Smart home

■ Smart grid

■ Intelligent traffic and transport system

■ Smart hospitals

■ e-governance

■ Smart offices

■ Smart water, gas distribution

■ Smart factory

■ Smart agriculture.

■ and many more...

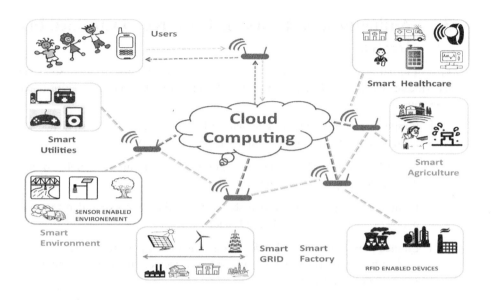

Figure 3.18: Cloud centric IoT.

As shown in Figure 3.18, all the smart services transmit data to the cloud storage via micro-processors and via gateways. So whenever the user wants to access the services, it has to register with cloud server and get the access for particular services. Major challenge for this approach is to identify authentic user and non authentic users. So identifying attacker is very difficult aspect because any user can register as a genuine user and later on try to apply denial of service attack using flooding approach. In the internet of things, major resources for the devices are battery and memory. So unnecessary requests or data transfer on any node can cause damage. We will discuss more challenges in next section.

Now in cloud computing, sometimes it is not possible to process all the data transmitted by each device, so fog devices do the local data processing using limited storage available to them and forwards to cloud. So cloud will have necessary information only. If the user wants to access specific services of any specific device then it will have to register with local server via fog device. This approach can reduce the burden on the cloud and help in the implementation of **distributed service support in IoT**. In the distributive service support environment, we can implement parallel cryptographic algorithms like light weight algorithms at local server and complex algorithms at cloud storage. Examples like biometric-based authentication can be resolved at local server rather than in cloud storage. It saves lots of network traffic and computation. So now let us discuss major challenges from which resource constrained IoT devices suffer.

3.5 Authentication Issues for Resource constrained IoT Devices

First of all let us discuss the parameters of resource constrained devices. So as per the Mocana, there are five major parameters that can prove that the device is resource constrained:

1. Limited memory in bytes or kilo bytes: Major resource constrained devices do not have built in memory or setup for external memory.

2. Limited processing: Major devices can process a very limited number of bits per second or bytes per second

3. Limited size: Size of the device should be as tiny as possible due to larger number of deployment and limited space coverage.

4. Limited power supply: Major devices are either self powered or battery powered so battery consumption should be lower. Device should work only when some processing is running.

5. Less flexibility: In major IoT devices, it is expected that we should not replace it at least for two to three years or more.

Major challenges faced by resource constrained devices are :

- Limited resources

- Weak network setup

- Secure light weight cryptography

- Lack of cyber awareness especially regarding secure credentials

- Upgrading devices with latest technology

Examples of some of the resource constrained devices given in [In (2012)],

- **Device type:** Crossbow TelosB, **CPU:** 16 bit MSP430, **RAM:**10 kBytes, **ROM:** 48 kBytes

- **Device type:** RedBee EconoTAG TelosB, **CPU:** 32-Bit MC13224, **RAM:**96 kbytes, **ROM:** 128 kbytes

- **Device type:** Atmel AVR Raven TelosB, **CPU:** 8-Bit ATMega1284P, **RAM:**16 kbytes, **ROM:** 128 kbytes

- **Device type:** Crossbow Mica2, **CPU:** 8-Bit ATMega 128L, **RAM:**4 kbytes, **ROM:** 128 kbytes

So implementation of authentication algorithm in resource constrained devices is a very difficult task due to complex algorithms and operations in cryptography. So researchers nowadays are focusing on providing equal security level using light weight operations. Another major challenge for the internet of things is to maintain the public key infrastructure or key distribution or certificate distribution center.

3.6 Application oriented authentication scenarios

Every application in the internet of things involves a large number of devices with different technologies, different protocols and different capabilities. Deploying this large number of devices is complex and protecting these devices is really a big challenge. Some of the applications like smart home where mobility of devices is low but threat from physical attack is high, are vulnerable. Some other applications like smart health where major devices are wearable and suffer from high mobility so threat from DoS type attack is greater. So in this section we will go through generalized scenarios of major internet of things applications and discuss the different places where implementation of authentication mechanisms must be deployed.

3.6.1 Health care

In the internet of things, researchers in major countries are focusing on health care importance due to its critical and need of emergency services. In the health care the major entities involved are:

■ Patient (with mobility or without mobility)

■ Doctor (either local or remote)

■ Hospital infrastructure

■ Ambulance

■ Wearable devices

■ Medical supply chain

■ Insurance companies

■ Government authorities

So as shown in Figure 3.19, a smart health care system can be deployed in distributed server based scenario for the larger scope and single server based scenario for local use like hospital or area. Major data collection in IoT health care will be through sensing enabled wearable devices. These major wearable devices will be enabled with RFID chip or bio-chip with unique identity like electric product code or printed bio-electro code. Other data collecting devices in health care are mobiles which can gather information like counting footsteps or walking speed and so on.

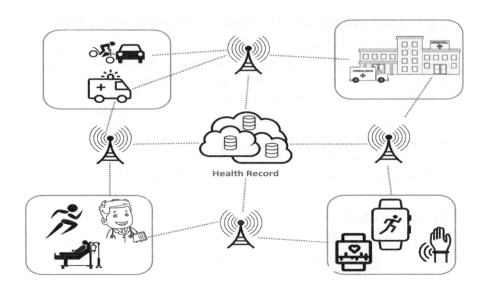

Figure 3.19: IoT Health care scenario.

So these all wearable devices and mobile devices communicate with health care infrastructure backbone, called gateways via some of the protocols like RFID, NFC, Zigbee, BLE, WIFI, 2G/3G/4G-LTE, Z-Wave and so on. Gateways collect the data and transmit this data to the servers. Hospitals, doctors, emergency services, medicine supply channel, health insurance companies can collect the data from the servers and provide the necessary services. So now let us try to understand the need for authentication: Major activities happening in patient data collection, patient data storage and patient data retrieving. Gateway and remote server play major roles in providing service, and these two points can be a honey pot for the attackers. Authentication in health care maintains security trio CIA as well as privacy and access control. Authentication in smart health care can be seen as:

- User authentication, user can be patient, doctor, medicine supplier, nurse or any other valid entity.

- Device authentication:
    - Authentication between wearable device and gateway
    - Authentication between mobile device and gateway
    - Authentication between gateway device and remote server
    - Authentication within medicine supply chain, which involves vehicles, production unit, research unit and so on.

Every user who wants to access the data for any purpose needs to register. During registration collecting role of user and attributes of user can help in access control mechanism. User registers with the servers and also gateway if needed. If gateway is working as a fog device then it also processes the data. Major challenges in smart health care is to get protection from fake patients and doctors, notorious wearable devices, fake emergency signals, unnecessary device usage. Wearable devices don't have higher power supply so continued access can affect the battery life. So identifying any user who pretended to be an authentic user and wants data from same device continuously without any specific reason is a major challenge. There are many other cyber challenges in IoT based smart health care. Authors in [Aslam et al. (2016)] have discussed authentication schemes for telecare medicine information system(TMIS). They have discussed entities for which two factor and three factor authentication is suitable in health care.

3.6.2 Manufacturing and logistics

Launching of industry 4.0 had opened the gate for a major revolution in the industry perception, It involves automation in manufacturing, smart service supply, smart raw material management, smart packaging, smart logistics and supply. Smart manufacturing involves various entities like

- Raw material management

- Employee management

- Product designing

- Product quality management

- Customer services

- Product packaging

- Product supply chain management

- Government and insurance agencies

- Plant management including plant security from fire and other disasters.

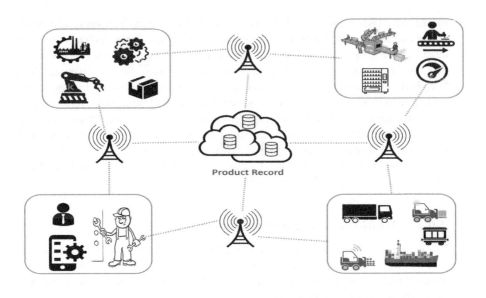

Figure 3.20: IoT Manufacturing scenario.

As shown in Figure 3.20, we can divide smart manufacturing into four parts, smart production, smart packaging, smart service and smart supply. So authentication, trust and access control is important. Different entities will have different levels of access. Local server collects the data about the plant and does the processing. All the employees working there will have to register. All the devices will be either RFID enabled or bluetooth or NFC enabled and will communicate with each other in a secured way but communication between gateway and server will be unsecured.

Multiple gateways deployed inside plant will continuously capture the data and send it to server. Users who want to control particular devices or machines need to get access via gateway. They need to register with the local server and generate the secret key to access. In smart logistics, containers will be deployed with various sensors that keep watch on the quality of the product and maintain the temperature. These sensors may not be internet enabled. With the help of a GPS device, users can keep watch on the location of product. If any accident occurs then with the help of smart road, it can immediately contact emergency service. All the IoT applications relate to each other when we try to deploy on the ground so smart infra development is important. Use of supervisory control and data acquisition system helped industry to monitor production using PLC and PID controllers. SCADA system provides control on local plant from remote location. But the major problem in the SCADA system is to protect the SCADA system. Proper authentication mechanism on SCADA system can help to protect from unauthorized gateway and denial of service attack.

3.6.3 Grid

The current grid system is one way communication in which consumers are passive in terms of energy generations and consumption requirement while smart grid provides two way communication where consumer can also generate renewable energy and can demand their power requirement to power distribution center. The power plant generates the power and transmits high voltage to power distribution center, power distribution center transmits power in low voltage to homes. In smart grid, current infrastructure needs to update where power transmission happens in two way. The major challenge for smart grid infrastructure is to authenticate the entities involved in a complete hierarchy. Proper authentication and access control mechanisms help to provide privacy of energy consumption and energy generation. Smart meters in the grid keep watch on energy consumption by the user. Smart grid infrastructure involves home infrastructure, area infrastructure, city infrastructure and power plant.

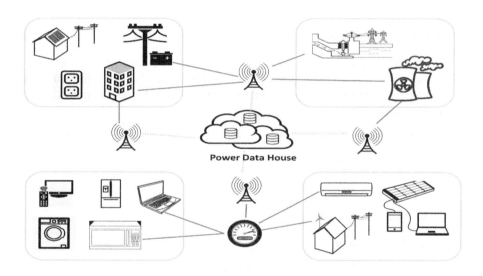

Figure 3.21: IoT electricity management scenario.

A home infrastructure is equipped with various various smart devices like smart lights, smart fridge, smart oven, smart washing machine and so on. This all the devices will be connected with a smart meter. In the initializing phase, all the devices will register with the smart meter. So whenever a new smart device becomes part of home network, it registers with the meter. Major communication protocol used here will be Zigbee. Authentication between a smart meter and home devices is important for protection from unauthorized device entry in to home network. A smart meter will collect data from each smart device on how much energy is consumed by each device, and which device needs more electricity. If the home is also enabled with a solar rooftop then switching between electricity by plant and electricity by solar can be managed with the help of a smart meter.

A major challenge will be maintaining privacy with devices available in the home and energy consumption by the home. Smart meters communicate with the gateways and provide details about the home network. This communication is most vulnerable due to its insecure communication via open protocols like wifi or ethernet. So we need strong authentication mechanisms between the smart meter and gateway. If an intruder can get access to smart meter or gateway then it can collect some private details about the home like number of devices and types of devices used by persons. So this needs strong security. Network of gateways make network of area and city, which finally stores these details into power data house storage from where power plant authorities, billing authorities, and other authorized entities can collect the data and apply the mining algorithm to collect more subjective and informative details. Figure 3.21 shows the basic entities involved in smart grid.

3.6.4   Agriculture

The agriculture industry is not explored that much in the internet of things application due to a more concentrated focus on health and business. Here we will discuss smart agriculture scenarios and various possible security issues. The major entities involved in smart agriculture are:

- Farmers and farms

- Fertilizer supplier

- Crop monitoring center

- Water supply center

- Electricity supply

- Product collection center(Food Mandi)

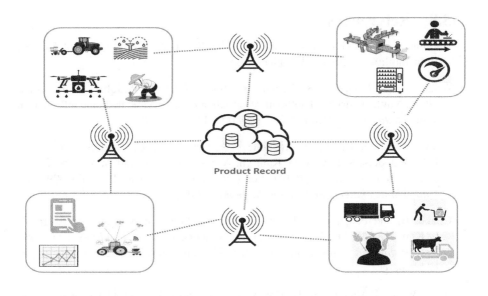

Figure 3.22: IoT Agriculture scenario.

Major security challenges are in maintaining communication between farmers and fertilize supplier, providing security to farm monitoring system, and handling government database. Other important parts of smart farming involve smart animal monitoring especially in the forest area. Smart forest monitoring involves care of animals and protection from fire.

3.7 Summary

In this chapter, we discussed a generalized authentication requirement and its phases involved. In the first section we discussed IoT authentication layered architecture and authentication requirements at each layer. In the second section, we discussed major types of devices involved and different authentication phases like registration, key exchange, login and authentication, password update, device adding phase. Cloud computing is the most important technology for internet of things, and major data will be stored in the cloud so we discussed a cloud centric approach for the internet of things and the need for authentication. In the last section, we discussed application wise security loopholes and authentication requirements in each application. We will discuss in more detail in subsequent chapters.

3.8 References

Abomhara, M. and G. M. Køien (2015). Cyber Security and the Internet of Things : Vulnerabilities , Threats , Intruders. *4*, 65–88.

Aslam, M. U., A. Derhab, K. Saleem, H. Abbas, M. Orgun, W. Iqbal, and B. Aslam (2016, Nov). A survey of authentication schemes in telecare medicine information systems. *Journal of Medical Systems 41*(1), 14.

Bliman, M., D. S. Bonale, N. Rochelle, J. A. Elwood, M. C. Hood, S. E. Isenberg, P. D. Saunders, S. L. City, D. Kathryn, A. Shah, R. John, and A. Mayers (2015). (12) Patent Application Publication (10) Pub . No .: US 2015 / 0073997 A1. *United States Patent Application Publication 1*(19).

Chang, H. and E. Choi (2011). User authentication in cloud computing. In T.-h. Kim, H. Adeli, R. J. Robles, and M. Balitanas (Eds.), *Ubiquitous Computing and Multimedia Applications*, Berlin, Heidelberg, pp. 338–342. Springer Berlin Heidelberg.

Chen, W., G. P. Hancke, K. E. Mayes, Y. Lien, and J. H. Chiu (2010, April). Nfc mobile transactions and authentication based on gsm network. In *2010 Second International Workshop on Near Field Communication*, pp. 83 89.

In, O. (2012). Management of Resource Constrained Devices in the Internet of Things. (December), 144–149.

Jones, A. K. and L. Lamport (1981). Password Authentication with Insecure Communication. *24*(11).

Li, N. and D. Liu (2017). Lightweight Mutual Authentication for IoT and Its Applications. *14*(8).

Mukherjee, M., R. Matam, L. Shu, L. Maglaras, M. A. Ferrag, N. Choudhury, and V. Kumar (2017). Security and privacy in fog computing: Challenges. *IEEE Access 5*, 19293–19304.

Schmitt, C., T. Kothmayr, W. Hu, and B. Stiller. *Two-Way Authentication for the Internet-of-Things*.

Shimizu, A. (1991). A dynamic password authentication method using a one-way function. *Systems and Computers in Japan 22*(7), 32–40.

Wang, B. and M. Ma (2012, Aug). A server independent authentication scheme for rfid systems. *IEEE Transactions on Industrial Informatics 8*(3), 689–696.

Yaga, D., P. Mell, N. Roby, and K. Scarfone (2018). Blockchain Technology Overview Blockchain Technology Overview. *National institute of standards and technology*, 1–57.

Chapter 4

Single Server Authentication

CONTENTS

✓The IoT is removing mundane, repetitive tasks or creating things that just weren't possible before, enabling more people to do more rewarding tasks and leaving the machines to do the repetitive jobs.

Grant Notman
Head of Sales and Marketing, Wood & Douglas

4.1 Abstract

Authentication in cryptography is the backbone that ensures confidentiality, integrity and availability. Certain applications may be running on a single server, and the user expects service from that single server. So a user needs to authenticate with the server to take the access. In this chapter we have discussed a single server scenario for authentication and the various phases involved in it.

4.2 Introduction

In the previous chapter, we went through various authentication scenarios by IoT devices in different IoT applications. Different authors have proposed different authentication schemes for the internet of things scenarios. Authors in [Aslam et al. (2017)] discussed various authentication schemes required in telecare medicine information system. In [Kothmayr et al. (2012)], the authors discussed two way authentication protocol implemented using datagram transport layer security, and the authors have highlighted authentication environment for resourceful devices and resource-less devices. For the resourceful devices like gateway or servers, it is possible to implement algorithms like RSA and AES, but for the devices that are resource constrained, they requires lightweight mathematical operation based algorithms like elliptic curve based cryptography. Authors in [Arasteh et al. (2016)] have proposed authentication schemes using three-factor environments. In their schemes, the user and sensor device need to register with the gateway nodes and gateway node work as intermediate server.

As an earlier discussion noted, radio frequency based identification will be the widely accepted identity mechanism for sensor devices. RFID will be the core technology for the device level communication in the internet of things. Many different authors have proposed authentication schemes for the RFID based device authentication. Authors in [He and Zeadally (2015)], have surveyed different authentication schemes based on RFID and elliptic curve cryptography for the health care environment. RFID Communication involves three major entities: RFID tag, RFID reader and gateway node. Communication between RFID tag and RFID reader and between RFID reader and gateway node must be secure enough. Most of the RFID based authentication schemes must ensure that need not be a third party required for authentication due to less resource and computation. So RFID authentication must ensure mutual authentication and other security parameters.

, In general, as shown in Figure 4.1, every authentication schemes in the internet of things environment must satisfy the following security requirements.

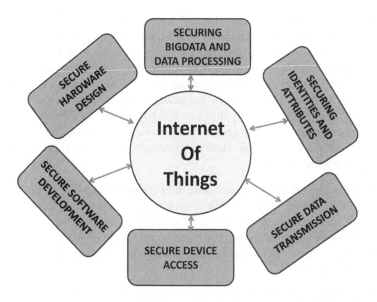

Figure 4.1: IoT Authentication.

■ **Securing identities and attributes:** An authentication scheme must protect the identity and attributes of the devices and uses, which are communicating with each other. If the identity of the devices is attacked, and the attacker is able to capture the identity of the communicating parties, then it can easily eavesdrop on some of the important information. For example, if the attacker can get the electronic product code for RFID tag then it can get information like type of product, name of company, quantity of product and so on. Similarly an attack on attributes can be easily done if the attacker continuously monitors types of signals, size of communication, and frequency of communication. For example, if the attacker can monitor electro-magnetic signals then he can capture useful information regarding the product from RFID tags. So an authentication scheme must ensure that it provides security for the identity and attributes of the device and users.

- **Securing data and control transmissions:** Every small bit of data and control transmission plays a crucial role in the internet of things environment due to high heterogeneity. Key established by the authentication scheme must ensure that all the communication transmitted by the key will provide enough confidentiality of data and control information.

- **Securing device access:** Access control mechanisms in applications like health care and smart industry play significant role. Authentication schemes based on attributes can be used for access control also, which is based on authentication identity of the user or device; they can then have access to sensors or data. Identity based authentication schemes can also provide mechanism for identity based access control. For example, if your identity is authenticated as an owner of the company then you will have access to the complete system of the company and if the identity is authenticated as an employee then you will have limited access. So authentication schemes can also be used for access control mechanism.

- **Securing big data storage and data processing:** Most of the applications in the internet of things will generate large volumes of data by sensing so we need to store this data in the local server also called as a gateway or fog device or any other server. The cloud server should be secured through a proper access control mechanism so it can be achieved using robust authentication mechanism.

- **Securing software development:** Most of the IoT data will be accessed by mobiles, so mobile applications should be developed in such way that they follows proper authentication mechanisms to communicate via gateway. It helps to secure application and provide user privacy.

- **Securing hardware design:** Most of the IoT hardware devices like sensors and microprocessors are not designed with built in security mechanisms but data generated by these devices sometimes may relieve device information so it need to protect using some security mechanism. With the help of proper authentication schemes we can secure it.

IoT applications like smart home, smart office, smart industry requires local data processing, and they need devices like local server or fog device or gateways, which can store and process the data. A single server environment ensures availability of server, which works as a inter mediator between user and sensors data, while in the multi server environment we will have one registration server or key distribution center. Multi server environment will be discussed in Chapter 5, while in this chapter, we will discuss single server scenarios in which gateway or fog device is considered as a server. IoT authentication major challenges are as follows:

■ Different types of devices manufactured by different industries without common standardization

■ Large variety in communication protocols followed by devices, Like RFID or NFC, BLE.

■ Mobility of users from one gateway to an other gateway

■ Dynamic topology

■ Different identity mechanisms

■ Resource constrained devices

■ Heterogeneity of involved entities

Every authentication protocol involves client-server model. In the internet of things, for the internet of things environment, a client can be user or device that wants to communicate with outside world via gateway. The server in the internet of things can be a gateway or cloud server that stores the user data and processes the data and also handles the authentication mechanisms like key exchange. In the next section, we will go through basic TCP client server model followed by data-gram transport layer security client server model and MQTT publish subscribe model. Both DTLS and MQTT are the most famous protocols used at network layer and application layer respectively in IoT.

4.3   Client-server Model

Hyper text transfer protocol is one of the most famous protocols used in the current internet. HTTP protocol follows client server model, and it is stateless protocol so it maintains cookies to support faster access of website. So basic client server model is as follows:

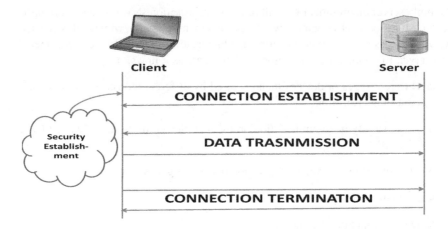

Figure 4.2: Client server basics.

As shown in Figure 4.2, the client-server architecture involves three major phases:

1. **Connection establishment:** Connection establishment phase establishes connection with the server. This step ensures exchange of required security parameters and authentication. After connection establishment, both client and server can have necessary security parameters like session key or digital certificate that both the entities can use for further communication. This phase may use a secure channel for some of the communications.

2. **Data transfer:** Data transfer phase makes use of security parameter and transmits the data on the insecure channel.

3. **Connection termination:** After successful data transmissions, connection termination phase destroys the session parameters and leaves the connection with server. It ensures both the entities have destroyed previous session parameters.

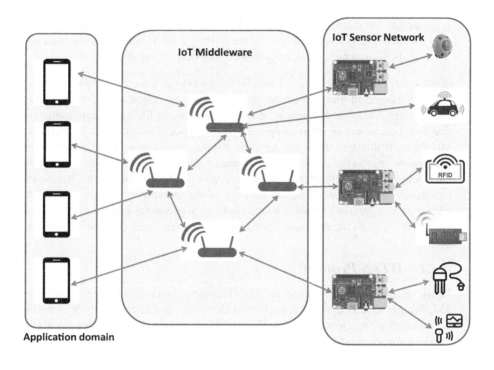

Figure 4.3: Basic IoT Client-server scenario.

As shown in Figure 4.3, internet of things generalized topology can be divided into three major domains:

1. **IoT Application domain:** Internet of things application domain consists of major services and mobile applications developed, which receive data from cloud server or local server. Application domain also consists of wearable devices that gets the signals based on recent activities as well as irrigation system, which behaves based on current moisture level. All of the IoT systems can be controlled using an application domain and all the applications (or users) need to register with servers to receive particular services.

2. **IoT Middle ware:** IoT Middle ware consists of gateways and fog devices. In some of the local applications like smart home, these gateways or fog devices work as servers. Every application domain application and device as well as sensor network processor and controller have to register with middle ware to get the services and provide services. IoT Middle ware is one of the most important part of the deployment of successful authentication mechanisms.

3. **IoT Sensor network:** Sensor network in the internet of things consists of very tiny and low capability sensor devices that collect the data from the environment or from the human body or from industrial materials and forwards this

data to the micro-processors(MP) for further storing and processing. IoT sensor network is a very soft part of the topology due to few resources and major challenges. It provides data collection and data aggregation facility.

In these three major domains, IoT middle ware works as a server in the single server IoT architecture. In this application, domains work as a client or user that wants to receive or to access the IoT sensor network. Applications like smart home, smart office, smart classroom or smart campus have these types of topological configurations. Mobile applications or users register with gateway to receive service(data) of particular sensor or group of sensors. Most of recent gateway devices and mobile devices are power supplied and capable enough to perform authentication tasks, but the major challenge lies in the IoT sensor network where sometimes microprocessors also want to receive data or control from the application and needs authentication. So, we need light-weight authentication algorithms for these resource constrained devices.

4.3.1 DTLS Protocol

In the internet we make use of TLS(Transport layer security) protocols and SSL(Secure Socket Layer) protocol to provide privacy and data integrity between two communicating applications in the client server environment. TLS makes use of symmetric cryptography for data encryption and public key cryptography for authentication purposes. To provide integrity, it makes use of MAC (Message Authentication code) function. But the major challenge in the TLS protocol is it makes use of TCP (Transmission control protocol), which is reliable but slow. It uses three-way hand shaking model for the communication, and it is very slow compared to UDP(User datagram Protocol). So DTLS (Datagram Transport Layer Security) protocol is derived making use of UDP as a transport layer protocol, which makes the connection less and faster protocol. There are many variants of DTLS protocol that can make use of RSA, Diffie-Hellman and ECC. Fully authenticated DTLS handshaking and key exchange in the DTLS protocol can be described as follows [Kothmayr et al. (2012)]:

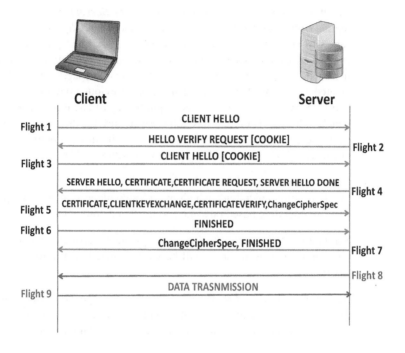

Figure 4.4: Datagram transport layer security.

Message sequences are called flight numbers and given sequences as occurrences.

1. Client sends a *ClientHello* message to the always running server. This message contains protocol version supported by the client and encryption algorithms it supports.

2. Server sends *ClientHello − Verify* message with cookie to check that client can receive as well as send the data.

3. Client replies with the *ClientHello* message cookie. These three steps ensures server that it is protected against some attacks like DoS.

4. Server sends a *ServerHello* message. *ServerHello* message contains half unprotected pre-master secret. This message contains the encryption algorithm, which is chosen by server and *ServerCertificate* to authenticate it self. With the same message, server sends request to client to provide certificate if client wants authentication for itself. *ServerHelloDone* message indicates end of communication for current sequence.

5. Client retrieves half pre-shared secret from *ServerHello* message. Client replies with *ClientCertificate* to authenticate itself. *ClientKeyExchange* message will be also transmitted, that contains another half pre-master secret encrypted by public key of server. So server can only retrieve this other half if it has correct private key. *CertificateVerify* message is hashed value calculated by private key of client using all previous communication that occurred. Server also verifies by using public key of client if both values match, which indicates that client has valid private key for the available public key. *ChangeCipherSpec* message indicates that all further communication from client will be via negotiated encryption algorithm and keys.

6. Client sends *Finished* message that contains encrypted message digest of all previous communications, which provides confidence to both the parties that they are operating based on same and unaltered communicated data.

7. Server also replies with *ChangeCipherSpec* message indicating that all further communication from server will be via negotiated encryption algorithm and keys.

8. Server sends *Finished* message that contains encrypted message digest of all previous communications, which provides confidence to both the parties that they are operating based on same and unaltered communicated data.

4.3.2   MQTT Protocol

Message queuing telemetry transport is an application layer protocol, which works based on a publish-subscribe mechanism. If client or user wants to receive data of particular sensor then user has to register or subscribe to the channel on which sensor publishes its values. MQTT protocol needs one broker who works as a intermediary between data generator and data receiver. Some of the famous MQTT brokers are RabbitMQ, Mosquitto or HiveMQ. MQTT is light weight, simple and easy to implement transport protocol. MQTT provides three types of quality of service message delivery:

1. **"At most once"** : This type of message delivery is suitable in an environment where if some data loss occurs or some individual text does not reach the user, then also it does not make high impact.

2. **"At least once"**: This type of message delivery is suitable for the environment where message must be delivered even if duplication occurs.

3. **"Exactly once"**: This type of message delivery is suitable for the environment where message must reach only once. Example is a payment system where message loss can deny the payment and duplication can repeat the payment.

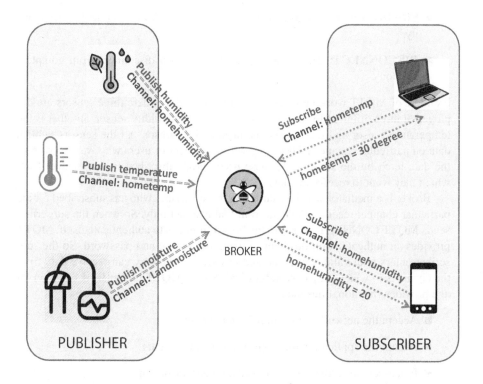

Figure 4.5: MQTT Publish-subscribe scenario.

Some of the major MQTT control packets are:

- **CONNECT:** This control packet is used by client to request a server for establishment of connection.

- **CONNACK:** This control packet is used for the acknowledgment of connection request.

- **PUBLISH:** This control packet is used to publish a message.

- **PUBAACK:** This control packet is used for publishing acknowledgment.

- **SUBSCRIBE:** This control packet is used to subscribe on particular channel.

- **SUBACK:** This control packet is used for providing acknowledgment of subscription.

- **UNSUBSCRIBE:** This control packet is used to unsubscribe from the particular channel.

- **PINGREQ:** This control packet is used to check availability of broker.

■ **PINGRESP:** This control packet is used to give acknowledgment of availability.

■ **DISCONNECT:** This control packet is used to disconnect from complete topology.

Example of MQTT working is discussed in Figure 4.5, where three sensors are deployed at three different locations. One sensor is the humidity sensor, an other is the temperature sensor and the last one is thr humidity sensor. All the sensors publish data on particular channels. There are two subscribers or users who want to receive the data fetch by the sensors. So these users have subscribed to the channel from which they want to receive the data.

Broker is a mediator which ensures that the subscriber who has subscribed for the particular channel receives the data from that channel only. So when the subscriber sends MQTT CONNECT packet to the broker, it needs to authenticate itself. MQTT provides an authentication mechanism using user-name and password. So the subscriber enters both valid user-name and password, then they can receive data from the channels to which they have subscribed. So the MQTT broker works as a server and provides functionalities like:

■ Accept the network connection from the clients

■ Accept the application messages published by clients

■ Processes subscribe and unsubscribe requests from client

■ Forwards application messages to the suitable clients

MQTT protocol suffers from some of major attacks like denial of service attack, injection of spoofed control packets, timing attack and so on. So MQTT protocol needs proper authentication mechanism, which provides valid communication. Some of the entities like NIST, FIPS, IEEE have made security mechanisms for MQTT communications. So MQTT implementation requires:

■ Authentication between user and device

■ Authorization of access to server resources

■ Integrity and privacy of MQTT control packets

Some of the other application layer protocols like CoAP are also used for IoT deployment, but we have limited the scope of discussion towards understanding client-server model and learning the two most widely deployed and used protocols DTLS and MQTT. Now for every attack, there will be an adversary who is either an individual or group of individuals who try to gain access to resources and try to perform some attacks. Some of the adversaries work for fun, while some work for particular gain. So let us see what adversary models are possible for IoT deployment.

4.4 Adversary Model

Deriving and identifying a simple, powerful and realistic adversary model is one of most important aspects of any cryptographic model. An adversary model provides an opportunity to the algorithm developer in such a way that he can identify the strength and weaknesses of an algorithm developed.

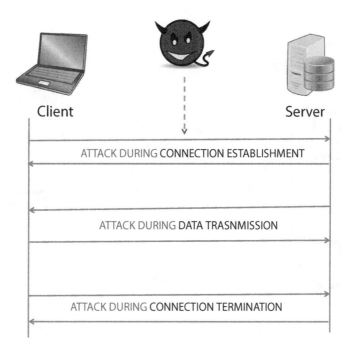

Figure 4.6: Adversary Basic scenario.

In general terms, if we want to define an adversary then we can say that "an adversary is any user who performs an active attack or passive attack by applying all the capabilities it has. " Adversary can be any user, it can be either registered user of the system or unregistered user of system. Registered adversary is more dangerous than unregistered adversary. Basic goal of an adversary is to extract the secret data from the communication and prevent or intercept the communication between authentic entities. As shown in Figure 4.6, adversary can attack during all the phases of communication.

An active adversary can attack during the connection establishment; during this phase important parameters like passwords and session keys are communicated over the channel so adversary will try to eavesdrop, modify, or block the message to perform attacks like denial of service attack, password grabbing or identity theft. Some of the authors like [Messai and Seba (2016)] have proposed an adversary model that can extract all the information from the compromised node, including secret keys, information and programs. Some of the schemes like [Messai and Seba (2016)] considered two types of adversary model.

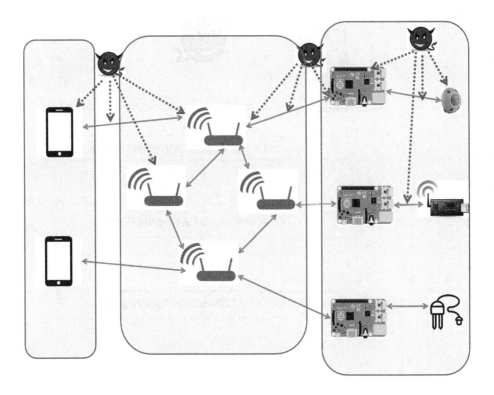

Figure 4.7: Adversary IoT scenarios.

1. **Temporary adversary model:** In this model, adversary tries to perform attack for some time with some pre-specified intention and goals.

2. **Permanent adversary model:** In this model, adversary continuously performs attack on the scheme and grabs the information or prevents the users from communication.

So an adversary is the one who tries to get unauthorized access of authenticated resources, or secret resources and most of the cryptographic schemes are monitored to be secure from attacks performed by an adversary. In the schemes where user uses smart card for secret communication, an adversary tries to steal the smart card. With the help of stolen smart card, he can retrieve important information like user id, time stamp, size of message, type of message. So he can easily track the devices and users. Adversary can block, delete, modify and reroute any message in the communication. Some of the adversaries continuously monitor the session key and try to make the session key for the next sessions. So as much as a strong adversary model an author keeps, he needs to generate as much secure algorithm.

As shown in Figure 4.7, adversary tries to perform attack on user nodes like identity theft or availability status, or it attacks on gateway node or server to capture the data and perform denial of service attack. Adversary can also have full access of public channels. In the internet of things scenario, a major challenge is providing security to devices from the physical theft by an adversary. Another major challenges is most of the industries who makes IoT devices do not consider the security important so any adversary who can just deploy the tools like wire shark can also capture the very useful information.

4.5   Phases

In single server based authentication schemes, we will have mainly five major entities: user, smart card reader, server or gateway, micro-processor and sensor. In the authentication scheme, we try to ensure that both the communicating parties are confident enough with each other about their identity and availability. There are five major phases that every successful authentication scheme needs to follow:

1. Server initialization phase

2. User or device registration phase

3. Login phase

4. Authentication phase

5. Password updation phase.

There can be other phases also like device revocation phase or deletion phase, but they may not be considered a part of authentication so it is beyond the scope of chapter.

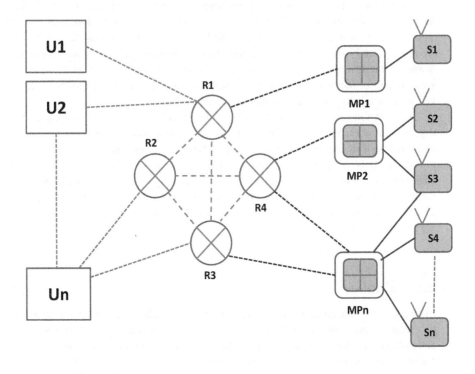

Figure 4.8: IoT Single server scenario.

As shown in Figure 4.8, internet of things single server architecture involves users $U_1, U_2, \ldots\ldots, U_n$. We assume that every user is enabled with the smart card reader program. It also involves gateway or routers $R_1, R_2, \ldots\ldots, R_n$ considered a server here. Micro controllers or micro processors are devices that work as a actuator, data aggregator and data forwarder. In major application scenarios, these devices will be communicating with the gateway and the gateway makes sure of the authenticity of these devices. So micro-processors $MP_1, MP_2, \ldots\ldots, MP_n$ works are inter-mediator between gateway and sensor in normal scenarios. Some of the sensors that are built in wi-fi enabled and are capable enough to directly communicate with gateways, then in that type of topological scenario, we do not need micro-processor type specific devices for data transmission.

Sensors $s_1, s_2, \ldots\ldots, s_n$ are devices that capture the data from the environment and transmit it. In the internet of things based authentication, a major challenge is to implement cryptographic algorithms on sensor devices due to their low capabilities. Most of the sensor devices forward data in the plain-text format so any adversary can easily capture the data. So now, let us discuss each phase for authentication in very generalized way for the internet of things. We consider it a generalized formulation because it may be possible that based on some different schemes and application, the author may change topology, but the order in which authentication will be carried out will be as follows.

4.5.1   Server initialization phase

We consider a gateway as a server device, and it is assumed that the gateway is always on the device. But there are some attacks in which an adversary may damage gateway devices, so we consider that an adversary that does not perform physical attack when gateway device is running. Server initialization is the first phase in authentication in which the server broadcasts all its secret parameters like public key or nonce values.

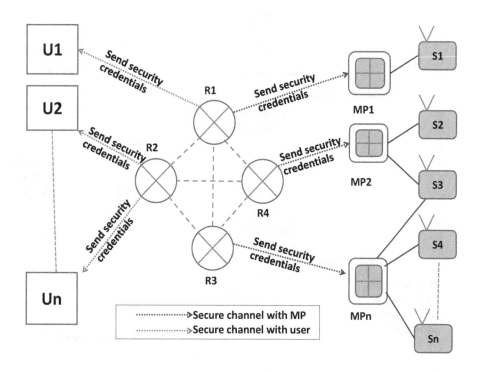

Figure 4.9: IoT Single server generalized initialization scenario.

So it depends on schemes prepared by the author whether or not the server sends these parameters using a private channel or public channel, but in major scenarios the server transmits public key type parameters publicly and generates values like nonce or seed that will be transmitted using a secure channel. As shown in Figure 4.9, server broadcasts secure parameter using a secure channel with all the micro-processors and users that are connected to it. This phase is optional in schemes where author considers that servers do not share any parameter before authentication. Every user and micro processor receives this parameter and do the computation as per security algorithms. Next phase is the user or device registration.

4.5.2   User registration phase

The user registration phase or device registration phase is considered an actual starting point of authentication. In client server communication, the client will initiate a connection with the server.

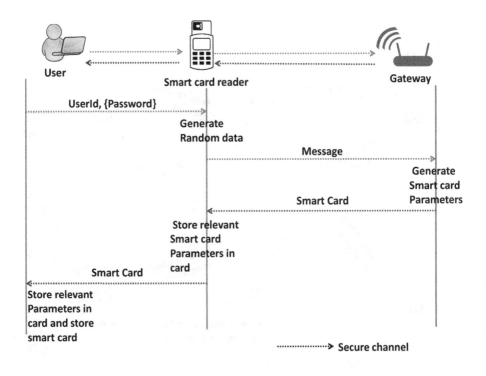

Figure 4.10: IoT Single server user registration generalize scenario.

As shown in Figure 4.10, in this phase there are three major entities that are involved, user, smart card reader and gateway. We have considered smart card availability in the authentication scheme. In some of the schemes, it may be possible that smart card based security schemes are not available, and the author shows mutual authentications. In this phase, message transmission will be as follows:

1. User will enter user id and password in the smart card application. In figure, password is shown in curly braces, which indicates there can be any type of password. Password can be either plain text or biometric traits or attributes.

2. Now smart card will generate any other random data and make a message to communicate with gateway(server) and sends it.

3. Gateway receives this message and performs necessary computations and generates the smart card parameters. It makes use of message sent by smart card reader, which is generated based on user id and password. Gateway generates smart card, stores it and forwards it to the smart card reader.

4. Smart card reader receives smart card from gateway(server) and stores any other relevant parameters in the smart card if required and sends smart card to the user or device.

5. User receives smart card and stores it with other relevant parameters required.

All the major communication in the registration phase is carried out in secure channels. User register ensures that gateway device is informed about user parameters and user can communicate using the parameters. Next phase is login phase.

4.5.3 Login phase

In login phase, user enters valid parameters and sends it to servers either in plain text or encrypted versions. For most of the current real time applications, user id and passwords are not transmitted in plain text. They are either encrypted or hashed using hash algorithms. Major communication in login phase onward will be in open channel. It is assumed that it is not possible to follow all the phases in the secured channel. As shown in Figure 4.11, message transmission in the login phase will be as follows:

1. User enters user id, password and smart card with the smart card reader.

2. Smart card reader verifies the user id, password and smart card.

3. If these values are not verified successfully, then it discards the session.

4. If these values are verified successfully, then it starts the authentication phase.

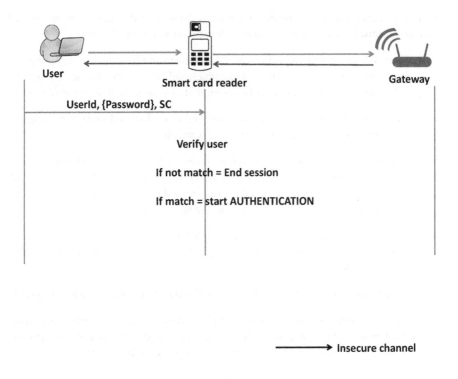

Figure 4.11: IoT Single server user login generalize scenario.

Successful computation of login phase does not ensure user will get services; every user needs to authenticate with the gateway so it can ensure gateway that user is valid user and has access to the services. So the next phase is authentication phase.

4.5.4 Authentication and key agreement phase

Authentication phase performs two major tasks:

■ Authenticate user or devices and validate its accesses

■ Generate session key for the current session and send this session key in such a way that no adversary can capture session key.

In the authentication phase, servers make use of users publicly available parameters and store parameters. Servers ensure that session key is only computable by valid user and any adversary, even knowing all the public parameters and previous session keys, can't calculate the session key. So as shown in Figure 4.12, in authentication phase, messages will be communicated as follows:

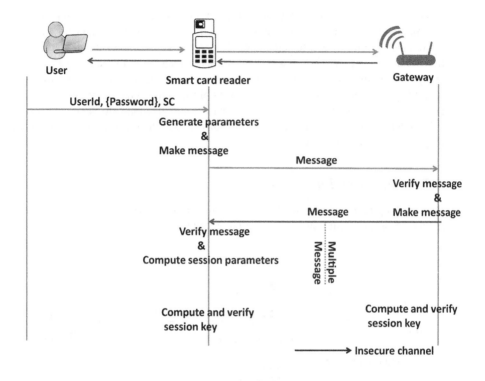

Figure 4.12: IoT Single server user authentication generalize scenario.

1. Smart card readers receive user id, password and smart card from the user.

2. It verifies it and if verification is successful, it generates parameters based on some random values (like time stamp), makes the message and sends it to server.

3. Server verifies the authenticity of message and user; if both the message and users are successfully authenticated, then it computes session parameters, store this session parameters in message and forwards this message to smart card readers.

4. Smart card reader verifies the message and computes session parameters. It is possible that multiple messages will be communicated between smart card reader and server before computation of session key.

5. Both smart card reader and server compute session key and optionally verify session key for both sides.

So after the authentication phase, server is ensured of identity of users and session keys $Sk_1, Sk_2, Sk_3, ..., Sk_n$ will be generated for n users. Both user and server will make use of this session key for the further communication. After the completion of communication, both the server and user destroy the session key, and for the next communication they need to regenerate the session key. Device authentication also follows the same procedure in which device sends its identity to the gateway devices and gateway devices make use of this identity to generate session key or temporary secret key to communicate with device. Next phase is the updating phase in which if user want to change its password or any device wants to change its security parameters. So let us discuss the password update phase in which user is motivated to update its password stored in smart card.

4.5.5 *Password updating phase*

Password updating phase involves communication between user and smart card reader. Message communication in the password phase occurs as follow:

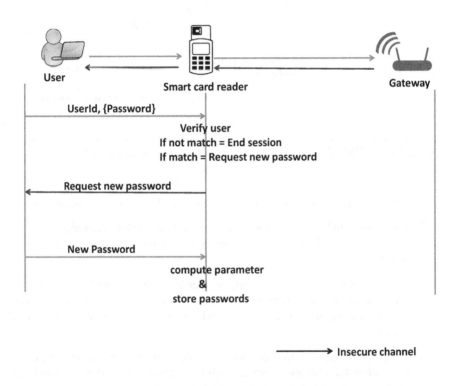

Figure 4.13: IoT Single server user password update generalize scenario.

1. User sends user id and password to the smart card

2. Smart card verifies the user id and password. If user id and password do not match, then it ends the session: if user id and password successfully verified, then it requests a new password

3. User receives the request and enters new password

4. User computes new parameters and sends it to smart card readers

5. Smart card reader computes new password using own parameter and stores it

So we have seen five major phases of single server based authentication schemes.

4.6 Summary

In this chapter, we have discussed the basics of authentication. We have gone through client-server model and IoT client-server scenarios. We have seen how the users can communicate using DTLS and MQTT protocol. MQTT protocol works on publish-subscribe based module. Where data collector publishes data on the particular channel and data user subscribes on that channel to make use of it. Later on in this chapter we discussed five major authentication steps that are necessary. So in general, this chapter gives a clear idea to new researchers who want to work on authentication and specifically in the IoT paradigm. They have learned, how they can prepare the single server based IoT authentication schemes.

4.7 References

Arasteh, S., S. F. Aghili, and H. Mala (2016, Sept). A new lightweight authentication and key agreement protocol for internet of things. In *2016 13th International Iranian Society of Cryptology Conference on Information Security and Cryptology (ISCISC)*, pp. 52–59.

Aslam, M. U., A. Derhab, K. Saleem, H. Abbas, M. Orgun, W. Iqbal, and B. Aslam (2017). A Survey of Authentication Schemes in Telecare Medicine Information Systems. *Journal of Medical Systems 41*(1).

He, D. and S. Zeadally (2015, Feb). An analysis of rfid authentication schemes for internet of things in healthcare environment using elliptic curve cryptography. *IEEE Internet of Things Journal 2*(1), 72–83.

Kothmayr, T., C. Schmitt, W. Hu, M. BrÃijnig, and G. Carle (2012, Oct). A dtls based end-to-end security architecture for the internet of things with two-way authentication. In *37th Annual IEEE Conference on Local Computer Networks - Workshops*, pp. 956–963.

Messai, M.-L. and H. Seba (2016). A survey of key management schemes in multi-phase wireless sensor networks. *Computer Networks 105*, 60–74.

Chapter 5

Multi-Server Authentication

CONTENTS

✓An easily accessible and transparent database of contract information will bring sunshine into the confusing and sometimes shadowy practice of government contracting.

Tom Coburn

5.1 Abstract

Various services like Google involve multiple services like YouTube, calender, finance and so on under a single roof. To use that service, users need to register one time with the google server and later on they can access all the servers using same user id and password. In this chapter, we discuss about various multi-server scenarios and phases involved in the multi-server environment. This chapter also discusses models of adversaries so readers will have a clear understanding about various loopholes possible in the multi-server environment.

5.2 Introduction

In Chapter 4, we discussed single server authentication scenarios and we also discussed phases involved in it. Single server authentication provides an environment in which user can decide its user id and password to access the server services. But after the iterative growth of internet and internet based services, the number of servers has increased. So in that circumstance, if we follow single server authentication scheme, then every individual server users needs to register a user id and password. For example, currently if we want to access Google and Yahoo service than we need a different user id and password, but if we want to access Gmail, Google calender, YouTube and many other Google services where each service has its own server then we can access that service using a single user id and password. So accessing multiple servers from remote access becomes difficult for remote users in single server authentication based schemes. Multi-server authentication provides an opportunity to access multiple services from a single login id and password. Formal definition of internet of things says providing service at any time to any one at any place. So it needs a distributed approach, which users can efficiently access. Some of the recent authentication schemes designed for multi-server environments are given in [Shunmuganathan et al. (2015)] [Luo et al. (2017)] [Chen et al. (2013)] [Chaudhry et al. (2015)] [Shingala et al. (2018)] [Amin et al. (2017)]

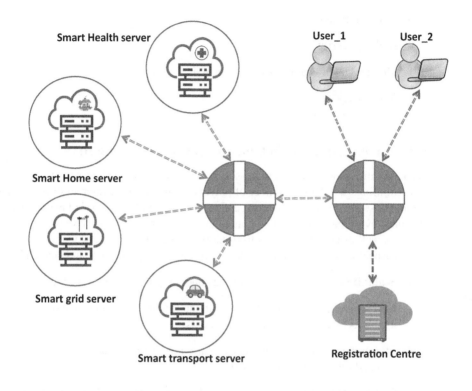

Figure 5.1: IoT Multi-server Environment.

As shown in Figure 5.1, various IoT applications like smart home, smart grid, smart health, smart transport and many more provide services, and they each have their own complex network. Every individual server and application will be connected with each other via different gateways and intermediate fog devices. A registration center is an entity with which every server and user is registered. After one time registration with the registration center, every user can directly access services of servers. Above topology is generalized topology for the internet of things environment, and it can vary for applications and requirements. For example, in larger applications like smart city, it may be possible that every city would have its own individual registration center with which every server running the city is registered. So all the people can access all the services of smart city using a single user id and password. Multi-server environment creates a smooth implementation of IoT environment, and multi-servers have the following advantages over single server authentication.

1. Easy and secure password modification

2. One time user and server registration

3. Less computation

4. Higher security

5. Mutual Authentication between client and server

6. Session key exchange

5.2.1 Challenges in IoT multi-server environment

The major difference between multi-server environment in the internet and in the internet of things will be the number of players involved. In the internet of things, the number of servers was less due to predefined services and locations but in the internet of things, every individual who wants to deploy smart home can have its own local server as well as needed services from global server also. So the scenario and topological aspect in the internet of things will be completely different and challenging.

1. **Scalability:** Number of local servers and gateways will be very high

2. **Heterogeneity:** Number of players in the IoT are many and each one is working on its own architecture and protocols

3. **Standardization:** After so many years, no commonly accepted standardization and reference model for IoT

4. **Security:**Huge number of servers will create open ground for adversaries ,

5. **Communication:**Implementing routing protocols will be very difficult in IoT multi-server environment

6. **Identification and authentication:**Authenticating valid client or user will be difficult due to heterogeneity

Multi-server environment will have many other challenges also like mobility, accessibility which we can consider the need to be focused. In this chapter, we will study multi-server environment and basic authentication phases that need to be considered. The basic scenario of multi-server based communication is as follows:

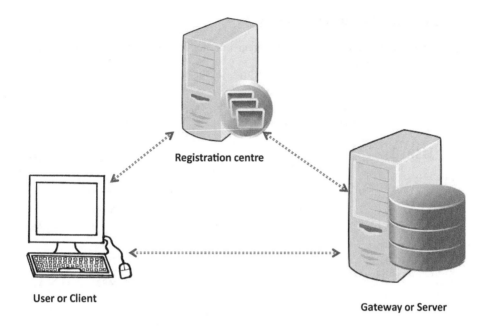

Registration centre

User or Client

Gateway or Server

Figure 5.2: IoT multi-server Basics.

As shown in the Figure 5.2, in multi-server environment, the major three entities will be involved in this: first one is client or user who wants to get service from multiple servers, an other is the server who provides different services and data, and another is registration center, that works as a trusted third party providing security services like keys exchange or smart cards for the secure communication between client and server. But a multi-server based scenario in the internet of things will be different, in which a client will require data from the server, and the server will ge t that data from different clients and smart devices that will generate that data. So as seen in Figure 5.2, $User_1$ wants data from the smart grid server, and it's possible that data is generated in real time by $User_2$. So in multi-server environment of internet of things, we can also say it is multi client - multi broker system in which multiple clients will publish data on multiple topics or channels and multiple brokers will receive and forward the data for subscribed users. But it generates the challenge of multi broker routing and mobility of clients. It is ensured that multi-server environment will create more attack vectors for adversaries.

5.3 Model of adversary

Model of adversaries in the multi-server internet of things environment we can understand using following:

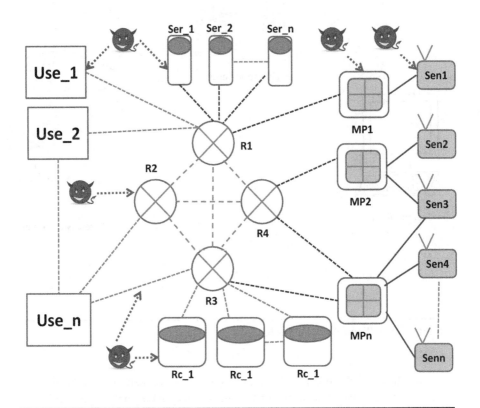

Figure 5.3: IoT multi-server adversary.

Multi-servers involve extra entities like registration center and different servers for different applications. Various attacks like distributed denial of service are a major challenge for this environment. As shown in Figure 5.3, adversaries can try to perform attacks on either user or devices or intermediate gateways, registration centers, micro processors and sensors. So adversaries model provides assumptions and advance knowledge about the attacker's capability. A strong adversary model will help us in the designing of strong authentication schemes. The basic threat models of an adversary are:

1. Adversary can be user, server, individual attacker or bots.

2. Adversaries will have full access of public channel so he/she/it can generate the

new message on a public channel, can delete and also can modify the messages over a public channel.

3. Adversaries are able to guess the less secured passwords, like user name it self a password.

4. Adversaries can steal the smart card and read the smart card, if it is transmitted on public channel in plain text format. With the help of sensitive data collection from smart card, an adversary can guess the other parameters like password and compute parameters like processing time and size.

5. Adversaries can obtain the old session keys then it may be possible that it can generate other session key.

6. Adversaries continuously monitoring link on registration center can have an idea about the number of users and servers.

7. Adversaries can perform physical attacks on sensors.

8. Adversaries can gain access to micro processors if they are not secured with secure transport-network layer protocols.

These are the various threat models possible in the internet of things multi-server environment. Attacks possible by adversaries in various scenarios are discussed in next chapter where we will study if adversary can capture some information, then what are the attacks it can perform.

5.4   Phases

multi-server environment for the internet of things can be seen as follows:

As shown in Figure 5.4, multi-server environment in the internet of things consists of four major clusters:

1. **User or client cluster:** In this cluster, all the users who want to access the data as well as who generates the data are part of it. Users who will generate the data will publish their data also.

2. **Gateway or Local server cluster:** In this cluster, all the intermediate gateway nodes and local servers are involved through which users are able to directly communicate and are capable of forwarding the user data to the users who need it.

3. **Sensor network cluster:** In this cluster, all the sensor and microprocessor devices involved will be there; in generalized IoT scenarios multi-processors do not do communication directly with registration center. Sometimes in the absence of routers, the user directly wants to access the data from microprocessor like raspberry pi, which is WiFi enabled then it needs to register for authentication and validation.

4. **Registration center:** Registration center can be a group of distributed centers according to application and requirements.

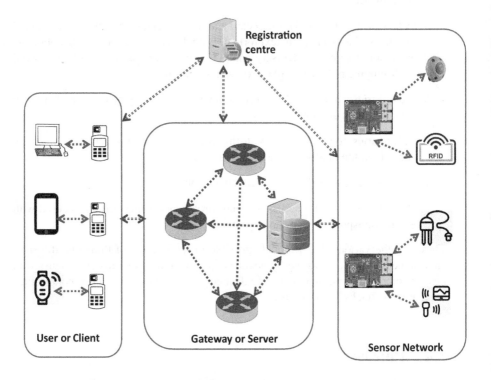

Figure 5.4: IoT multi-server.

In general for the algorithmic terms, the multi-server scenario can be seen as follows:

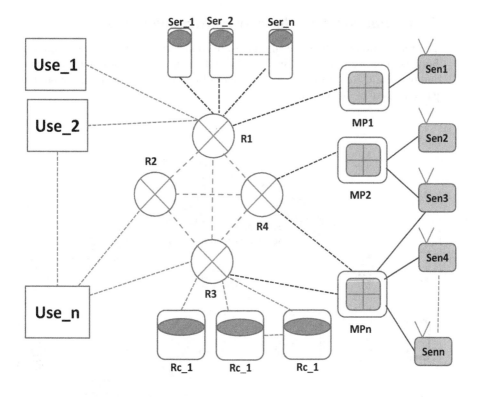

Figure 5.5: IoT multi-server devices.

In Figure 5.5, it shows that there are n users $(Use_1, Use_2, ..., Use_n)$ who want to get the service from n servers $(Ser_1, Ser_2, ..., Ser_n)$ and travels through n routers $(R1, R2, ..., Rn)$. Every n servers will register with one of the registration centers $(Rc_1, Rc_2, ..., Rc_n)$. Microprocessors $(MP1, MP2, ..., MPn)$ will collect the data from sensors $(Sen1, Sen2, ..., Senn)$ and publish. multi-server authentication schemes contains following phases:

1. **Registration center initialization**

2. **Server registration**

3. **Client registration**

4. **User Login**

5. **User authentication**

6. **User password update**

5.4.1 Initialization phase by registration center

The registration server is the first phase of multi-server authentication algorithms. Initialization is an optional phase in the authentication schemes and it depends on authors. Basic aim of this phase is to alert all the participating entities about public parameters of registration center.

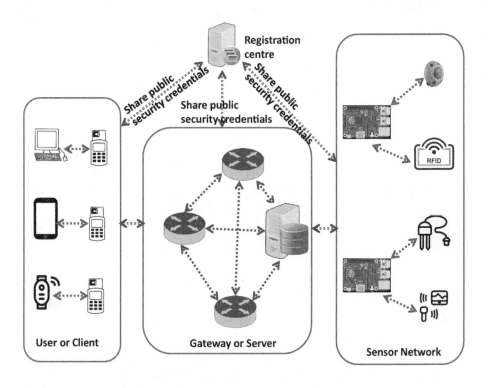

Figure 5.6: multi-server: registration center initialization.

As shown in Figure 5.6, every registration center publicizes its public key parameters and other values with all the participating entities. Every new user or server who wants to register with the registration center makes use of these parameters and registers with the RC.

5.4.2 Server registration

The server registration phase indicates that every server in the multi-server environment does the registration with registration center. During this registration process,

RC verifies the services provided by the server and assigns the secret parameters or smart card to the server. Complete communication of server registration will be take place under the secure channel. During preparation of this scheme, we need to ensure that it should use less computation and complex algorithms.

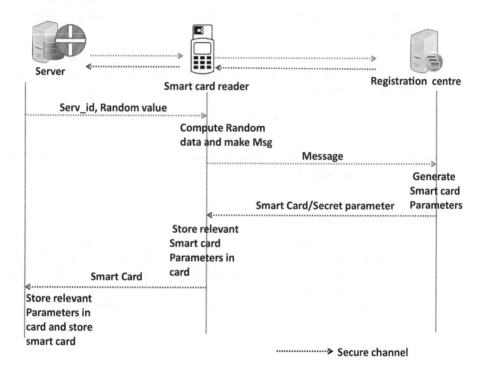

Figure 5.7: multi-server: server Registration.

As shown in Figure 5.7, there are three major entities involved in server registration. The server can be a local server or sometimes gateway/fog device in the IoT environment. The other entity is the smart card reader; smart card reader can be any application or actual smart card reader that accepts the server id and secrets required to identify server and complete further processing. And last entity is the registration center, which provides necessary secrets or smart card to the server for communication. Message passing in the server registration will be as follows:

1. Server will pass server id and secrets in the smart card application. Secrets can be any generated values that the smart card reader can verify.

2. Now smart card will generate any other random data and make a message to communicate with registration center and send it on secure channel.

3. Registration center receives this message and performs necessary computations and generates the smart card parameters. It makes use of message sent by smart card reader, which is generated based on server id and secret parameters. Registration center generates smart card, stores it and forwards it to the smart card reader.

4. Smart card reader receives smart card from registration center and stores any other relevant parameters in the smart card if required and sends smart card to the server or gateway.

5. Server receives smart card and stores it with other relevant parameters required.

In multi-server environment both client and server will do registration with RC, so next phase is user registration.

5.4.3   User registration

User registration phase involves the user who wants to access the data from server, a smart card reader and a registration center. In the IoT, protocols like MQTT provide an opportunity to client also for data generation so in that cases we can consider that the client will upload that data to the server and from the other clients will make use of it. Message passing in the user registration will take place as follows:

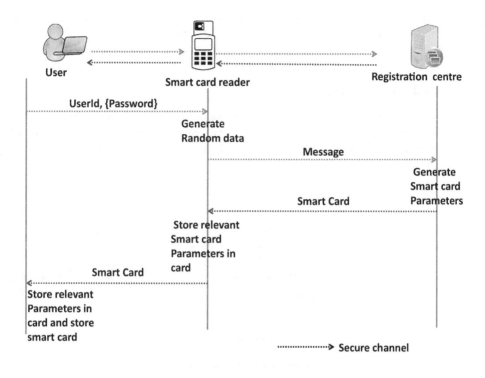

Figure 5.8: multi-server: user Registration.

1. User will enter user id and password in the smart card application. In the above figure, the password is shown in curly braces, which indicates there can be any type of password. Password can be either plain text or biometric traits or attributes.

2. Now smart card will generate any other random data and make a message to communicate with registration center and send it.

3. Registration center receives this message and performs necessary computations and generates the smart card parameters. It makes use of message sent by smart card reader, which is generated based on user id and password. Registration center generates smart card, stores it and forwards it to the smart card reader.

4. Smart card reader receives smart card from registration and stores any other relevant parameters in the smart card if required and sends smart card to the user or device.

5. User receives smart card and stores it with other relevant parameters required.

User registration phase ensures that user i will have smart card, which can be used for communication with any server j. After this phase user will have single smart card, which can be used for multiple servers registered with registration center.

5.4.4 *User login*

User login phase in the multi-server environment is similar to single server environment. The smart card reader will verify user id and password for the client. Password can be either text password or biometric. Message passing in the login phase will be as follows.

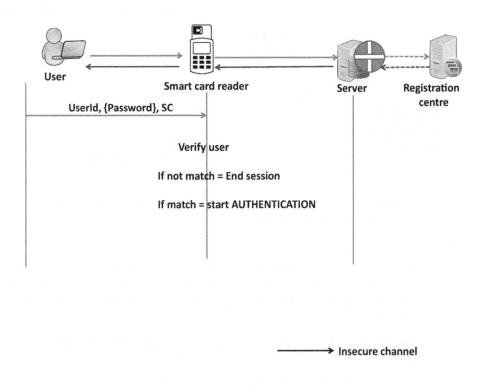

Figure 5.9: multi-server: user login.

As shown in Figure 5.9,

1. User enters user id, password and smart card with the smart card reader.

2. Smart card reader verifies the user id, password and smart card.

3. If these values are not verified successfully then it discards the session.

4. If these values are verified successfully then it starts the authentication phase.

After the login phase, the smart card reader will start authentication phase. Login phase does not ensure that server will provide service. User and server will perform mutual authentication to ensure identity of each other. Next phase will be user authentication phase.

5.4.5 User authentication and key exchange

User authentication phase ensures the server that user is valid user and ensures user that it is communicating with correct server. In multi-server, we make sure that server already has valid smart card or secret parameter provided by registration center, so it can make use of it for faster validation. Message passing in authentication will take place as follows:

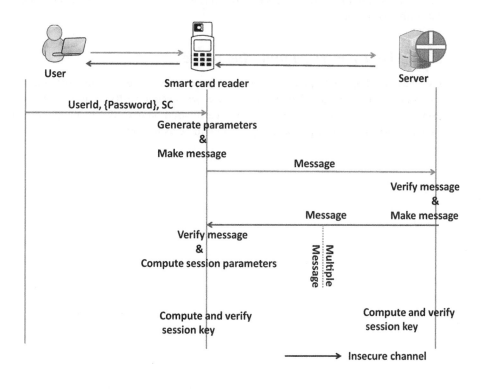

Figure 5.10: IoT multi-server : user authentication.

1. Smart card readers receive user id, password and smart card from the user.

2. It verifies it and if verification is successful, it generates parameters based on some random values like nonce, makes the message and sends it to server.

3. Server verifies the authenticity of message and user; if both the message and user is successfully authenticated then it computes session parameters, store these session parameters in message and forwards this message to smart card readers.

4. Smart card reader verifies the message and computes session parameters. It is possible that multiple messages will be communicated between smart card reader and server before computation of session key.

5. Both smart card reader and server compute session key and optionally verifies session key on both the sides.

After the successful authentication phase, both the client and server will have session key for the communication. So both of the entities will make use of that session key for further communication.

5.4.6   *User password updating phase*

User password updating phase in the multi-server environment is the most complex environment. In password updating phase, user will need to verify itself with previous user id, and later on need to get new smart card based on new user id and password. Message passing in user password updating will be as follows:

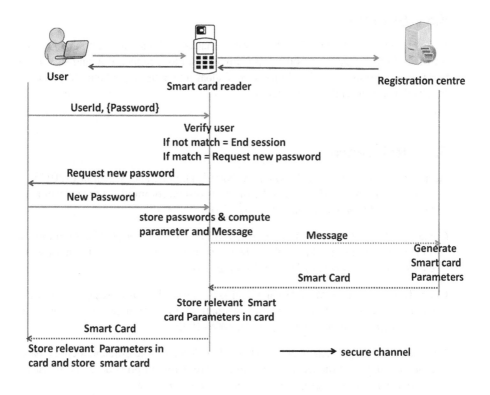

Figure 5.11: IoT multi-server : user password update.

1. User sends user id and password to the smart card

2. Smart card verifies the user id and password. If user id and password do not match then it ends the session. If user id and password are successfully verified then it requests a new password

3. User receives the request and enters new password

4. User computes new parameters and sends it to smart card readers

5. Smart card reader computes new password using own parameter, stores it and sends message with new parameters and old smart card to registration to get a new smart card.

6. Registration center will receive message and verify it. If verification is done successfully then registration center will compute smart card parameters and send to smart card reader.

7. Smart card reader gets new smart card and stores it after adding relevant parameters and stores it to user device.

5.5 Summary

In this chapter we discussed multi-server scenarios and authentication environment for them. In this chapter various phases involved in user registration and server registration are discussed in detail providing a clear understanding about all the phases involved in end-to-end authentication.

5.6 References

Amin, R., S. K. Islam, M. K. Khan, A. Karati, D. Giri, and S. Kumari (2017). A two-factor RSA-based robust authentication system for multiserver environments. *Security and Communication Networks 2017.*

Chaudhry, S. A., H. Naqvi, M. S. Farash, T. Shon, and M. Sher (2015). An improved and robust biometrics-based three factor authentication scheme for multiserver environments. *Journal of Supercomputing*, 1–17.

Chen, T.-Y., C.-C. Lee, M.-S. Hwang, and J.-K. Jan (2013). Towards secure and efficient user authentication scheme using smart card for multi-server environments. *The Journal of Supercomputing 66*(2), 1008–1032.

Luo, M., Y. Zhang, M. Khan, and D. He (2017). A secure and efficient identity-based mutual authentication scheme with smart card using elliptic curve cryptography. *International Journal of Communication Systems* (February).

Shingala, M., C. Patel, and N. Doshi (2018, Mar). An improve three factor remote user authentication scheme using smart card. *Wireless Personal Communications 99*(1), 227–251.

Shunmuganathan, S., R. D. Saravanan, and Y. Palanichamy (2015). Secure and efficient smart-card-based remote user authentication scheme for multiserver environment. *Canadian Journal of Electrical and Computer Engineering 38*(1), 20–30.

Chapter 6

Attacks and Remedies

CONTENTS

√The single biggest existential threat that's out there, I think, is cyber.

Michael Mullen

6.1 Abstract

Attack is the attempt to damage the system, and remedy is the way, we can prevent that attack, either fully or partially. During the designing of the scheme, if the researcher is aware about the various possible attacks and remedies then it can help during scheme designing. In this chapter, we discuss various attacks that most frequently occurs on any authentication schemes; and this chapter also provides various remedies for those attacks.

6.2 Introduction

Attack is an attempt to damage something. Attack in the world of internet is considered a cyber attacks. Cyber attackers can be an individual or a group of people. The motive of an attack can be either ethical or non-ethical. Ethical attackers are those who work for the betterment of people or country. In the internet of things, as discussed in Chapter 1, an attacker will have a large area for to attack. From device to user, they will have as big an attack vector as possible. There are many different types of attack possible but in cryptography there are six basic attacks that lead to the major possible attacks.

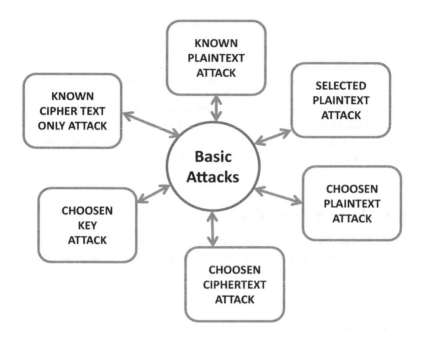

Figure 6.1: Basic attacks.

As shown in Figure 6.1, the following are basic attacks.

- **Known cipher text only attack:** In this attack, attacker will have access to cipher texts for each communication. Based on available cipher text, an attacker can derive plain text.

- **Known plain text only attack:** In this attack, attacker will have access to cipher text and some of the plain text like id. Based on available cipher text and plain text, an attacker will try to retrieve keys to encrypt message or algorithm to decrypt upcoming messages.

- **Selected plain text only attack:** In this attack, attacker will have access to cipher text and some of the plain text like id. Based on available cipher text and plain text, An attacker will try to select the plain text that is encrypted.

- **Chosen plain text only attack:** In this attack, attacker will have access to cipher text and some of the plain text like ID, based on available cipher text and plain text, an attacker will try to choose plain text and encrypt it using an available algorithms.

- **Chosen cipher text attack:** In this attack, an attacker will try to choose cipher text and try to decrypt using an available algorithm to match with the decrypted value.

■ **Chosen key attack:** In this attack, an attacker chooses a random key from the dictionary and tries to decrypt the retrieved cipher text.

All of these attacks provide basic understanding about available scenario of attack vector. Now let us discuss some famous attacks and their possible remedies in cryptography. In the internet of things, to provide better authentication schemes, researchers need to keep in mind all of these attacks.

6.3   User Anonymity

User anonymity can be defined as an "Attack to capture identity of the user with the wrong intentions". If an attacker can get the identity of the user during any phase or message of communication then it can:

■ Misuse identity to perform other attacks

■ Track the users and its activities

■ Break the privacy measurements

■ Pretend to be an authentic user

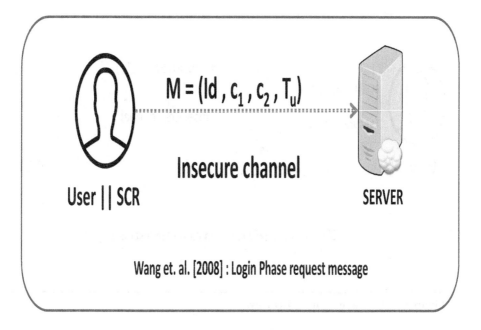

$$M = (Id, c_1, c_2, T_u)$$

Insecure channel

User || SCR

SERVER

Wang et. al. [2008] : Login Phase request message

Figure 6.2: User anonymity : Attack space.

User anonymity ensures that during any communication message, the identity of user should not be released. User anonymity provides privacy and liberty to user. Here we have taken examples from proposed schemes [Wang et al. (2007)]. Schemes proposed by Wang et. al. are not secured in terms of user anonymity. As shown in the Figure, during log in request, communication smart card reader sends *ID* and *ID$_i$* in plain text format over public channel respectively. As shown in Figure 6.2 any intruder or attacker can capture identity and misuse it. Remedies for user anonymity, are:

■ Propose a scheme in which users don not share its identity in plain text with the server during any phase of communication

■ Make sure that if user sends his/her user id, then it should not be in plain text format

To give an example, we have taken the scheme proposed in [Shingala et al. (2018)]:

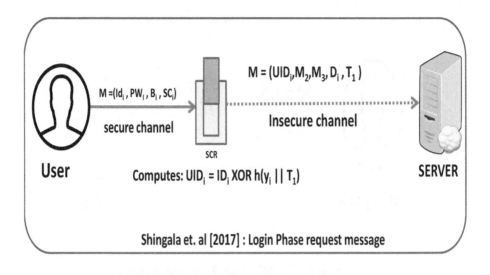

Figure 6.3: User anonymity : Remedy.

As shown in Figure 6.3, before sending identity, the Shingala et al. smart card reader computes: $UID_i = ID_i \bigoplus h(y_i||T_1)$ and sends UID_i as an identity for the login request. We have not discussed complete scheme here, but we want to show that identity of user will be transmitted after a certain operation on it, which will make sure that the attacker can not obtain the identity of user. The only server with the certain parameters available can verify the identity, but can not obtain the original identity.

In the internet of things, maximum communication will be between devices, so devices, will communicate with each other using their identities. During device authentication, device will need to prove that it is a valid device that can take the data or store the data or can publish the data and subscribe to the channel. During all these communications, authentication schemes that ensures device identity should not send identity of device in the public channel. Device anonymity break can help the attacker to track the device if it is a mobile device or obtain the other information like type of device, use of device, frequency of device data collection, energy consumption by device and so on. Ensuring device anonymity is also an important parameter for the internet of things security.

6.4   Perfect forward secrecy

Forward secrecy can be defined as "protecting the earlier history if the current security is broken". Perfect forward secrecy is the mechanism that ensures that if the server is attacked and compromised then the entire system of communication will not be null and void. So a scheme with perfect forward secrecy makes sure that even when the long-term secret key of the server is attacked and attacker obtains, it, then attacker must not:

1. get the older session keys

2. decrypt the old communication

3. compute the next session keys

4. encrypt the plain text in a similar way authenticated user is encrypting

Maintaining perfect forward secrecy is an important parameter of foolproof authentication. Some of the algorithms like Diffie-Hellman algorithm ensures perfect forward secrecy. To discuss an attack scenario, we have taken a message communication scheme [Wen and Li (2012)]. Here we will not discuss the complete scheme, but we will discuss what the missing point is in the scheme, which makes this scheme not secured from the perfect forward scheme. We can get details of complete scheme [Wen and Li (2012)]. In this scheme, shown in Figure 6.4 server chooses long-term secret key x, and it makes use of x for the computation of smart card (SC) and session key (SK), for the computation of session key server and user uses variables (A_i, T, B_i, T'). All these variables are derived from the static secret x. The major problem related to perfect forward secrecy here is that the server does not use any variable value that will be different for computation of new session key. We can say that this scheme is not secured from perfect forward secrecy because if the attacker who has continuously monitored all the communication and is able to retrieve the server secret x anyway, then he/she can compute all old session keys and can decrypt all past records.

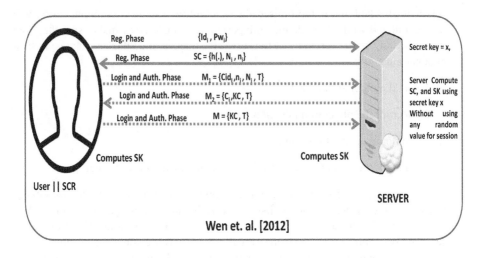

Figure 6.4: Perfect forward secrecy : Attack space.

Even if server or client does not get information about client then attacker can perform other attacks like chosen plain text attack. To show the remedy of this attack, we have proposed a very simple solution (may not be secured from other attacks).

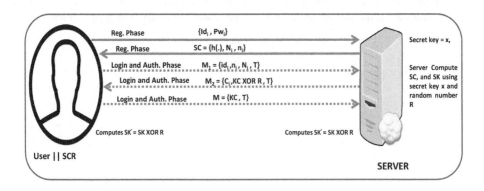

Figure 6.5: Perfect forward secrecy : Attack space.

As shown in Figure 6.5, server makes use of a random number for the computation of session key, which we believe will provide security for perfect forward secrecy. As a remedy of perfect forward secrecy attack, we can say that it makes use of some random numbers at both sides, for each new session in such a way that even an attacker gets the long-term secret key, he/she can never compromise older session keys and older messages.

6.5   Replay attack

Replay attack can be defined as an attack in which attacker may send the older message repeatedly to the server or client to misguide them. So whenever attacker replays same messages multiple times with the intention to perform the other attacks like denial of service attack, battery draining attack and so on, then we can say it is a replay attack. In the internet of things whenever attacker makes use of a replay attack with the aim of reduction of battery, it is the biggest challenge to stop the attacker from performing that. Here, we will consider that the attacker is capable of capturing the encrypted messages and can replay the same messages to the client and server time after time. To get an understanding of replay attack, we have taken a message communication from the scheme proposed in [Truong et al. (2017)]. Here we have not discussed the complete scheme because we want to get the understanding of a replay attack and to show a point that provides an opportunity to the attacker so that he/she can perform a replay attack.

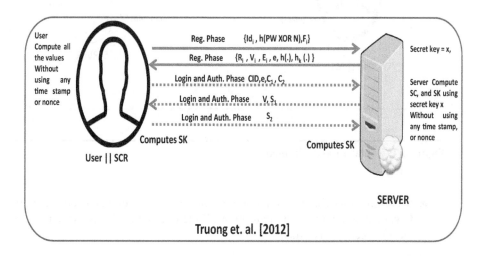

Figure 6.6: Replay attack : Attack space.

As shown in Figure 6.6, user and server do the following communication.

- User $\rightarrow$ Server on secure channel : $Id_i, h(PW \oplus N), F_i$

- Server $\rightarrow$ User on secure channel : $R_i, V_i, E_i, e, h(.), h_k(.)$

- User $\rightarrow$ Server on insecure channel : CID, e, C_1, C_2

- Server $\rightarrow$ User on insecure channel : V, S_1

- User $\rightarrow$ Server on insecure channel : S_2.

After this communication, both server and user will compute the session key. Now we can be the observer in this scheme that neither user side nor the server side, there is no mechanism implemented that makes the client aware. The server is not aware either of the arrival time of message or what might disable the complete possibility of replay of message. As a remedy of a replay attack, shown in Figure 6.7, we believe:

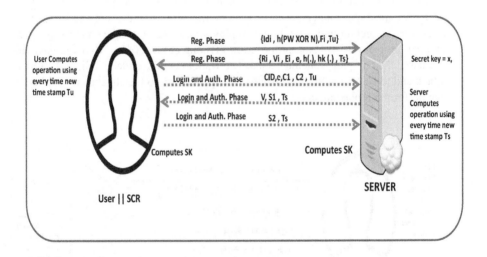

Figure 6.7: Replay attack : Remedy.

- Server and user can use time stamp for the communication and verify the difference between two communications, if the threshold value is exceeded then message can be discarded.

- Server and user can use random number during each message communication, which discards the possibility of repetition of same message.

- Server and user can use nonce and store the nonce value for each session, which will make each message different and neglect the probability of repetition of message.

6.6 Off-line password guessing attack

Off line password guessing attack is the most dangerous attack if an attacker is able to implement it within a polynomial time on the system. Successful implementation of off line password guessing attack opens complete system in front of attacker. Possibility of successful off line password guessing attack is increased in the scenario when an attacker is able to find a password dictionary suitable to system environment. In off line password guessing attack, an attacker continuously monitors all communicated messages and tries to gain maximum information from them using spoofing tools. To discuss the off line password guessing attack, we have observed a scheme proposed in [Li (2013)]. Here we will not go through complete scheme, but we will discuss necessary portion of it related to password guessing attack.

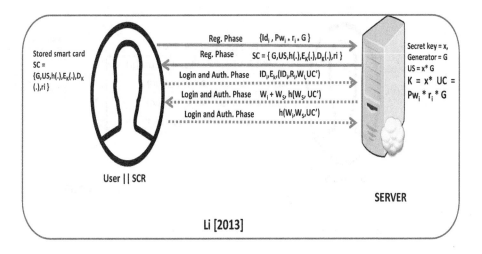

Figure 6.8: Off line Password guessing attack : Attack space.

As shown in the Figure 6.8,

- User → Server on secure channel : $(Id_i, Pw_i * r_i * G)$

- Server → User on secure channel : $SC = G, US, h(.), E_K(.), D_K(.), r_i$

- User → Server on insecure channel : $ID_i, Ek(ID_i, R_i, W_i, UC)$

- Server → user on insecure channel : $W_i + WS, h(WS, UC)$

Here we assumed that an adversary can steal the smart card and can read the smart card so the adversary will have $SC = G, US, h(.), E_K(.), D_K(.), r_i$ and have password guessing dictionary DG. When ever user sends a login request, he/she encrypts the communication using Ek. Server can obtain this key by computing $US = x * G$, where x is server secret and G is shared generator. Key $K = x * Pw_i * r_i * G.$ and can decrypt user login message. Now let us agree that the adversary has guessed password $Pw_{i'}$ and has smart card information. Adversary will be able to compute $K' = PW_{i'} * r_i * G$. Now using this K', adversary will try to decrypt log in message and compare the obtained ID' with the ID contained in login message. If $ID = ID'$ then we can say that guessed password $PW_{i'}$ is the correct password or else adversary can guess an other password and do the same operation again. Here it depends on certain assumptions, but we can understand that if maximum information is leaked during the communication then it increases the probability of successful password guessing attack. Figure 6.9 shows offline password guessing attack remedy.

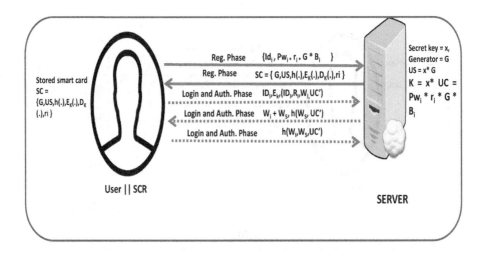

Figure 6.9: Off line Password guessing attack : Remedy.

As far as remedy concerns for the offline password guessing attack go, we believe that schemes should be developed after making use of,

■ Long-term secret key

■ Biometric based secret

■ Private random number for both sides

Most of the smart card based schemes provide protection from password guessing attack, but we need to keep in mind the complexity of scheme and computation parameters required for it. It should not make schemes on paper possible but in real time implementation of it may not be useful.

6.7 User impersonation attack

User impersonation attack is an attack in which attacker tries to obtain the identity of any valid communicating parties and communicate with an other communicating party using that. In an impersonation attack, the attacker tries to fool the communicating parties by pretending to be legitimate entity. Attacker tries to get access to various resources at the server side. It is a very similar attack such as when we have the password of some lab then we enter that lab and try to get access to resources. User impersonation attack is also very essential attack that needs to be taken care. There are so many schemes in which authors have successfully proposed user impersonation attacks and that attacks are proposed using different computations. There is no predefined parameter that if this is the scenario then user impersonation attack will be delivered but to discuss the user impersonation attack we make use of the unsecured scheme given in [Wang et al. (2009)].

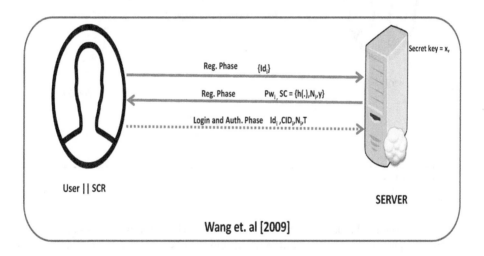

Figure 6.10: User impersonation attack : Attack space.

As shown in Figure 6.10, in this scheme the author sends login request via insecure channel that contains ID_i, CID_i, N_i and T. To discuss the attack, we will review the attack given in [Sood (2012)]. Let attacker intercept a login message then attacker uses current time stamp T_c and computes:

- DT $= T \oplus T_c$, $Ni' = Ni \oplus DT and CID'_i = CID_i \oplus DT$.

- Sends ID_i, CID'_i, NI', T_c

- Server matches time stamp and verify the computed identity ID'_i with ID_i.

As shown in [Sood (2012)], server can easily get that both the identities are similar. Server will pretend that it has received a message from the valid user. There are certain reasons why server is not able to identify that the user is not correct user. First reason is user communicate its ID in plain text so it makes an attacker easy to set parameters according to that. Another reason is for both sides, there is no random parameter during computation. Server makes use of constant secret x for the verification.

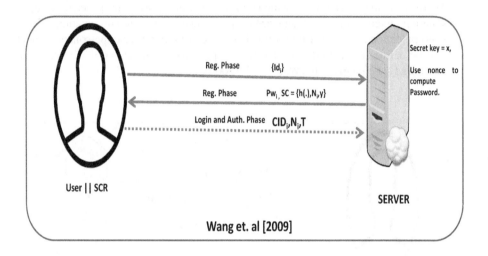

Figure 6.11: User impersonation attack : Remedy.

As shown in Figure 6.11 introducing remote variable for the computation protects from impersonation attack. And in the scheme user should never send its identity in plain text format.

6.8 Man in the Middle attack

Man in the middle attack is one of the most famous attacks in cryptography in which an attacker can capture the messages, can read the messages, can delete the messages and can alter the messages. In man in the middle attack, attacker alters a message in such a way that both the communicating parties think that they are communicating with the correct counterpart. Let us assume that Alice and Bob are communicating with each other.

■ Alice sends a message to Bob that "I will meet you today", but it is intercepted by Eve and sends to Bob that "I will not meet you today".

■ Bob sends message to Alice that "Ok, we will meet tomorrow," but it is intercepted by Eve and sends to Alice that "Ok, we will meet today".

In this conversation Alice thinks that Bob will come today to meet, but in reality Bob is going to meet tomorrow. So intercepter Eve has played the role of Man-in-the-Middle(MITM) here. Man in the middle attack is shown in 6.12.

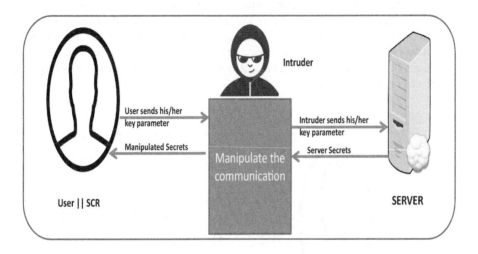

Figure 6.12: Man in the middle attack : Attack Space.

Sometimes in MITM attack, attacker separately communicates with both entities as a intruder. Using the remedy of the above communication, we can rewrite the communication with the help of one way hash function. As discussed in Chapter 2, due to its resistance proof properties, if an attacker changes any value then the communicating entities can easily capture it. For a remedy for man in the middle attack, we can use hash function, check sum, or message authentication code. A foolproof authentication mechanism ensures that the scheme is secured from Man-in-middle attack. Sometimes if an attacker is not an active attacker but he/she is a passive attacker then to intercept the messages, they make use of spoofing tools like Wireshark. With the help of wire shark or dsniff, an attacker captures the information about networks, types of data transfer, time of data transfer.

6.9   Smart card loss and stolen attack

It may possible that the user loses his/her smart card and the attacker gets it or attacker stole the user's smart card. After getting the smart-card based on power consumption mentioned in [Kocher et al. (1999)] and reverse engineering mentioned in [Messerges et al. (1999)], an attacker can capture the useful information stored in the smart card. After capturing the information from the smart card, the attacker may try to retrieve the password of user to perform other attacks like user impersonation and masquerading access attack or attacker tries to retrieve the long-term secret of a server. Stolen smart card attacks become dangerous when any of the above said parameters are calculated based on information stored in it. So every researcher who is preparing the scheme for the authentication based on smart card should take care of this. To discuss stolen smart cards or loss attacks we have taken message communication proposed in [Liu et al. (2008)] (for the shake of ease of understanding scenario). An attack is observed from the [Sood (2012)]. Here we have not discussed the complete schemes, but we just have taken a message which can help us to understand the stolen smart card attack.

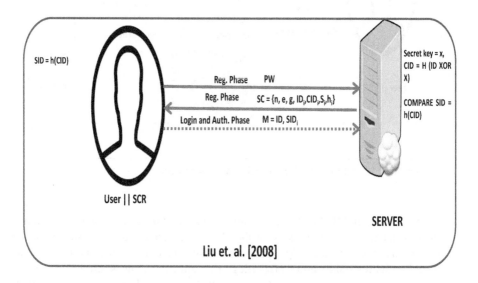

Figure 6.13: Smart card stolen attack: Attack Space.

As shown in Figure 6.13, the server sends a smart card $(n, e, g, ID_i, CID_i, S_i, h_i)$ here whenever user gives password to the server password chooses ID for the user and computes $CID = h(ID \oplus x)$ where x is the secret of server. Now let us assume that an attacker obtained the smart card from the user and gains the smart card information. Now user computes $SID = h(CID)$ and sends it for to login.

Similarly now an attacker will also have CID and ID so attacker will also compute $h(CID)$ and try to login with server. Server will be easily cheated and masquerade by the attacker. To provide a protection from this type of attack, an author should make use of some parameters that are only available with the server and client and those parameters may used to get an access to resources and used during an authentication phase. To get the solution as shown in Figure 6.12, we have proposed a very simple solution for the above said attack, if server performs the following operation:

■ $CID = h(ID \oplus x \oplus PW)$

■ $SID' = h(CID \oplus PW)$

■ $CompareSID = SID'$

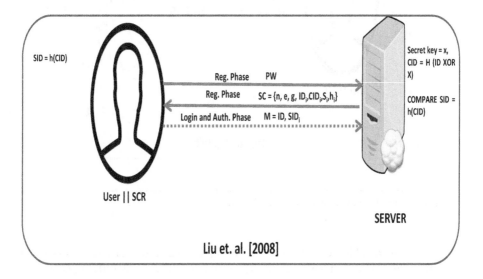

Figure 6.14: Smart card stolen attack : Remedy.

A smart card stolen attack as shown in Figure 6.14 can also lead to some dangerous attack like MITM attack or denial of service attack. Authors need to take care that smart card should not reveal identity, password or long-term secret.

6.10   Server spoofing attack

In a server spoofing attack, an attacker will create a fake server and will try to contact the users. Basic motive behind performing server spoofing attack is to get the information from the user and masquerade to the other servers. In an authentication scheme, there is an equal importance of server authentication to user authentication.

Mutual authentication between server and client is an important parameter. Server spoofing attack are sometime performed by a legal server to masquerade another server to get the data of the client. Let us see the scenario.

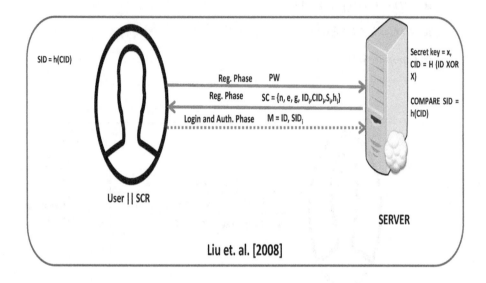

Figure 6.15: Smart card stolen attack : Remedy.

As shown in Figure 6.15, there are two servers that have registered with the same registration center and have a common secret like in [Lee et al. (2011)]. Now whenever any legal user tries to interact with any legal server, a server behaving as a masquerading server intercepts this message and generates the smart card in such a way that the user cannot understand that the smart card he received is not from the actual server he was expecting. Foe remedy concerns, in the multi server environment every server should have a unique secret that makes every user login with a different server with different parameters.

6.11   Denial of Service attack and Distributed DoS

Denial of service attack is the most famous attack in cryptographic history. In denial of service attack, an attackers motivation is to deny for the victims either temporary timing or permanent. An attacker can perform a denial of service attack by many different methods such as:

■ Attacker gets access of service and modifies the verification parameter

- Attacker sends continuous requests on the service

- Attacker physically damages the server or device

- Attacker stops the communication service

So these are the various possibilities in which an attacker can perform the denial of service attack. In the internet of things scenario, there will be a maximum number of resource constrained devices involved during communication, so there is a possibility that an attacker tries to attack physically or in terms of resource consumption. Attacker performs attacks as shown in Figure 6.16.

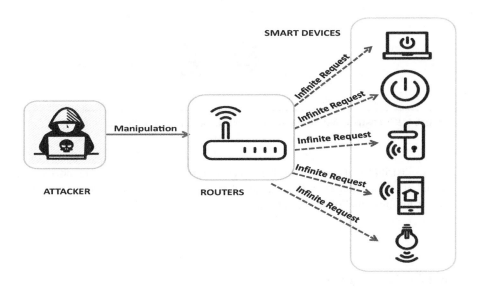

Figure 6.16: Denial of service attack : Attack scenario.

Some of the famous denial of service attack methodology is TCP syn flooding attack. In TCP, SYN is used for synchronization in the connection establishment, and it indicates that SYN requester wants to establish a connection. So an attacker manipulates a gateway in such a way that the gateway generates an infinite number of SYN requests. In denial of service attack, attacker tries to compromise one system or one device at a time. While in the distributed denial of service attack, an attacker compromises multiple gateways and using multiple gateways the attacker tries to damage the distributed system. Distributed denial of service attacks can be understood through Figure 6.17:

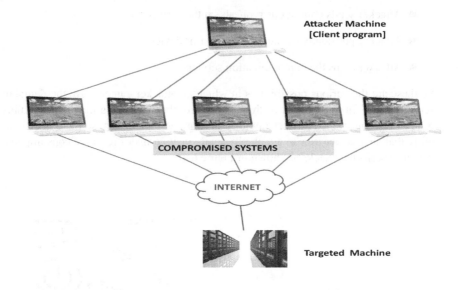

Figure 6.17: Distributed denial of service attack : Attack scenario.

So distributed denial of service attack is a much more dangerous attack than denial of service attack. In the internet of things, an attacker will have a big attack vector for the denial of service attack.

6.12   Summary

In this chapter, we discussed various attack scenarios possible in the internet of things. There are other cyber attacks that also exist like side channel attack, stuxnet and many more but we have limited ourselves to authentication based attacks. We discussed each attack with the attack space and we have given a scenario in which attack can take place. We have also given a remedy for an attack in which we discussed possible solutions for the attacks.

6.13   References

Kocher, P., J. Jaffe, and B. Jun (1999). Differential power analysis. In M. Wiener (Ed.), *Advances in Cryptology — CRYPTO' 99*, Berlin, Heidelberg, pp. 388–397. Springer Berlin Heidelberg.

Lee, C.-C., T.-H. Lin, and R.-X. Chang (2011). A secure dynamic id based remote user authentication scheme for multi-server environment using smart cards. *Expert Systems With Applications 38*(11), 13863–13870.

Li, C. T. (2013, March). A new password authentication and user anonymity scheme based on elliptic curve cryptography and smart card. *IET Information Security 7*(1), 3–10.

Liu, J.-Y., A.-M. Zhou, and M.-X. Gao (2008). A new mutual authentication scheme based on nonce and smart cards. *Computer Communications 31*(10), 2205 – 2209.

Messerges, T. S., E. A. Dabbish, and R. H. Sloan (1999). Power analysis attacks of modular exponentiation in smartcards. In Ç. K. Koç and C. Paar (Eds.), *Cryptographic Hardware and Embedded Systems*, Berlin, Heidelberg, pp. 144–157. Springer Berlin Heidelberg.

Shingala, M., C. Patel, and N. Doshi (2018, Mar). An improve three factor remote user authentication scheme using smart card. *Wireless Personal Communications 99*(1), 227–251.

Sood, S. K. (2012). An improved and secure smart card based dynamic identity authentication protocol. *International Journal of Network Security 14*(1), 39–46.

Truong, T. T., M. T. Tran, A. D. Duong, and I. Echizen (2017). Provable Identity Based User Authentication Scheme on ECC in Multi-server Environment. *Wireless Personal Communications 95*(3), 2785–2801.

Wang, X.-M., W.-F. Zhang, J.-S. Zhang, and M. K. Khan (2007, July). Cryptanalysis and improvement on two efficient remote user authentication scheme using smart cards. *Comput. Stand. Interfaces 29*(5), 507–512.

Wang, Y.-y., J.-y. Liu, F.-x. Xiao, and J. Dan (2009, March). A more efficient and secure dynamic id-based remote user authentication scheme. *Comput. Commun. 32*(4), 583–585.

Wen, F. and X. Li (2012). An improved dynamic id-based remote user authentication with key agreement scheme. *Computers & Electrical Engineering 38*(2), 381 – 387.

Chapter 7

Analytical Matrices and Tools

CONTENTS

✓ If security were all that mattered, computers would never be turned on, let alone hooked into a network with literally millions of potential intruders:

Dan Farmer

7.1 Abstract

Time complexity computes time required for the algorithm during run time. Space complexity computes memory consumption, and energy consumption helps to identify energy consumption. In this chapter, we discuss these three parameters, which can help evaluate algorithm asymptotic capabilities. Analytical tools like AVISPA, ProVerif and BAN-Logic help researchers validate security protocol designed by him/her and verify whether any attacks are available or not. In this chapter, we also discuss these tools and the working of these tools.

7.2 Analytical Matrices

Algorithm analysis provides understanding about behavior of an algorithm in terms resource consumption. In the internet of things, the major challenge lies in the development of communication protocols that make use of lower bandwidth, create low collision, generate law latency and provide better reliability. Similarly cryptographic algorithms can be analyzed by how many properties they satisfy, what the security level is, and how many resources they consume. For the internet of things enabled devices, major challenge will be fewer resources available for it. Internet of things enabled devices will suffer from less memory, less computation power and responsibility of quick response. There are three major parameters that we will discuss here.

■ Time complexity

■ Space complexity

■ Energy consumption

So let us discuss each parameter and understand the cryptographic algorithm from the resource point of view.

7.2.1 Time complexity

Time complexity can be defined as total time required in all the phases of authentication. As seen in Chapter 4 and Chapter 5, authentication involves the following phases:

■ Registration phase

■ Login and authentication phase

■ Password change phase

All of these phases involve certain operations. To measure time for these operations a common measurement is T_X shows that time required to compute the operation X. In cryptography authentication, some of the operations that are involved and the time measurement units for that operations are shown below.

■ T_h:Time required to perform single hash function

■ T_s: Time required to perform symmetric encryption/decryption operation

■ T_{enc}: Time required to perform message encoding operation

■ T_{dec}: Time required to perform message decoding operation

■ T_{mod}: Time required to perform modular operation

■ T_{xr}: Time required to perform Ex-OR operation

■ T_{pm}: Time required to perform point multiplication

Table 7.1: Time Consumption chart.

Author	Total cost(Registration phase+ Login/Authentication phase)
[Moon et al. (2017)]	$16T_h + 1T_H + 4T_{pm} + 2T_s$
[Nikooghadam et al. (2017)]	$8T_h + 7T_s + 2T_{xr}$
[Quan et al. (2017)]	$15T_h + 7T_{mode} + 1T_m$
[Truong et al. (2017)]	$15T_h + 5T_{pm} + 6T_{pa}$
[Amin et al. (2017)]	$12T_h + 4T_s + 4T_{mode}$

■ T_{pa}: Time required to perform point addition operation

■ T_{minv}: Time required to perform modular inverse

■ T_{mode}: Time required to perform modular exponentiation operation

- T_{PR}: Time required to perform pseudo random function

- T_{ecc}: Time required to perform complete elliptic curve operations

- T_{ch}: Time required to perform chebyshev map operations

- T_H: Time required to perform bio-hash function

Based on the total number of the time particular operations occurred, we can compare the authentication algorithms. A recent survey given in [Aslam et al. (2017)] has given computation cost for various authentication schemes proposed by different authors as shown in Table 7.1.

7.2.2   Space complexity

Space complexity is the total space required by the user and server to store the parameters for the computation of various phases. Space complexity of the link during transmission must be considered during the designing of the parameter. Algorithms must be designed in such a way that they requires a smaller number of message transmissions with a smaller number of parameters. Designed authentication schemes must be formalized in such a way so that a smaller number of parameters the user and server side need to store for the computation. If we compare two algorithms in a uni server environment where one algorithm uses biometric and other does not make use of biometric then the algorithm with biometric operation needs more space consumption due to the higher number of bits required to store biometric images in RAM during run time and ROM during off line time. So let us consider S_X as a unit of storage, which indicates size of parameter X required to store it at user side,server side or on link. Then in authentication algorithm we can have following space requirements(can be more than listed).

- S_{SC}: Space required to store smart card

- S_{ID}: Space required to store identity of user

Table 7.2: Space Consumption chart.

Algorithm	Key Size	Input size	Output Size
DES	56 bit	64 bit	64 bit
3DES	168/112/56 bit	64 bit	64 bit
IDEA	128 bit	64 bit	64 bit
CAST	40-128 bit	64 bit	64 bit
AES	128/192/256 bit	128 bit	128 bit
RC2	8-1024 bit	64 bit	64 bit
RC4	40-2048 bit	2064 bit	2064 bit
RC5	0-2040 bit	32/64/128 bit	32/64/128 bit
Blow Fish	32-448 bit	64 bit	64 bit
RSA	1024-4096 bit	-	-
DSA	1024 bit	-	-
ECDSA	163 bit	-	-
DH	1024 bit	-	-
ECDSA	163 bit	-	-
MD2	-	-	128 bit
MD4	32 bit	512 bit	128 bit
MD5	32 bit	512 bit	128 bit
SHA	32 bit	512 bit	160 bit
SHA1	32 bit	512 bit	160 bit

Table 7.2 shows space consumption by various cryptographic algorithms in bits.

- S_{PW}: Space required to store password

- S_{EC}: Space required to store elliptic curve

- S_{ECG}: Space required to store elliptic curve generator

- S_P: Space required to store elliptic curve base point

- S_{PR}: Space required to store random number

- S_{S_k}: Space required to store session key

- S_{Pr_k}: Space required to store private key

- S_{Pu_k}: Space required to store public key

- S_H: Space required to store bio hash

- S_{bio}: Space required to store extra computed parameter

7.2.3 Energy Consumption

Energy consumption can be defined as how much total energy is consumed during cryptographic algorithms and operations. We have taken scenario from [Potlapally et al. (2003)]. Measurement of energy consumption will be in three units, mJ (Milli Joule), μJ (Micro Joule) and MJ (Mega Joule). So let us see the energy consumption by different symmetric and asymmetric cryptographic algorithms.

Table 7.3: Energy Consumption chart. [Potlapally et al. (2003)]

Algorithm	Key generation	Enc/Dec	sign/verify	Key Exchange	Hash Generation
DES	27.5(μJ)	2.08(μJ/B)	-	-	-
3DES	87.04(μJ)	6.04(μJ/B)	-	-	-
IDEA	7.96(μJ)	1.47(μJ/B)	-	-	-
CAST	37.63(μJ)	1.47(μJ/B)	-	-	-
AES	7.87(μJ)	1.21(μJ/B)	-	-	-
RC2	32.94(μJ)	1.73(μJ/B)	-	-	-
RC4	95.97(μJ)	3.93(μJ/B)	-	-	-
RC5	66.54(μJ)	0.79(μJ/B)	-	-	-
Blow Fish	3166.3(μJ)	0.81(μJ/B)	-	-	-
RSA(1024)	270.13(mJ)	-	546.5/15.97	-	-
DSA(1024)	293.20(mJ)	-	313.6/338.02	-	-
ECDSA(163)	226.65(mJ)	-	134.2/196.23	-	-
DH(1024)	875.96(mJ)	-	-	1046.5(mJ)	-
ECDSA(163)	276.70(mJ)	-	-	163.5(mj)	-
MD2	-	-	-	-	4.12(μJ/B)
MD4	-	-	-	-	0.52(μJ/B)
MD5	-	-	-	-	0.59(μJ/B)
SHA	-	-	-	-	0.75(μJ/B)
SHA1	-	-	-	-	0.76(μJ/B)
HMAC	-	-	-	-	1.16(μJ/B)

So as shown in Table 7.3, major encryption algorithms require higher energy consumption compared to hash functions. In IoT based authentication schemes, hash function can be the better option for designing schemes.

7.3 Analytical tools

After the invention of internet various internet bodies like IEEE, IETF, W3C started to develop protocols for communication between two computers and between computers and servers. At the initial stage of development, there was no need of security, but after the expansion of internet, the attacker community also emerged and started to attack the systems and the messages transmitted. After various attacks, the research community started to develop cryptographic algorithms to transmit messages in cipher text rather than plain text. With the enhancement in cryptographic protocols, including IETF, NIST and IEEE, other individual researchers also started to develop security protocols, which provide security from various attacks and communicating parties can communicate securely. During the development phase, to check whether a developed algorithm is secured against various attacks or not, developer community required tools that can check whether a developed protocol is secured or not.

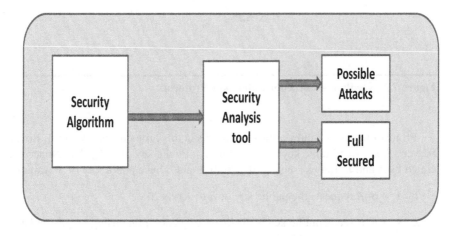

Figure 7.1: Cryptography protocol verification.

There are many different tools developed to verify security level of developed protocols. In this section we discuss most used security protocol analysis tools. As shown in Figure 7.1, every tool takes input as a developed protocol and verifies it. As output it gives whether algorithm is secured or not. If it is not secured then what are the different attacks that are possible and how is it possible. We will discuss AVISPA, Scyther, ProVerif, HERMES and BANLogic. We will discuss how tools work and what are the inner functions that make one better than another. For the verification of tools, we will take examples of the Needham-Schroeder public key protocol (NSPK). NSPK protocol can be understood by following Figure 7.2.

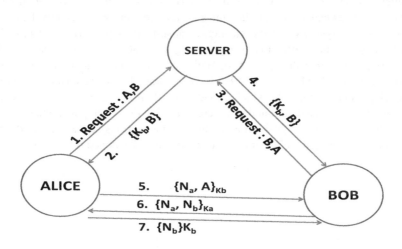

Figure 7.2: Needham-Schroeder Public key Protocol.

Protocol works as follows, Basic notations in this protocols are Alice: A, Bob: B, Server : S, K_a, K_b : Public keys of A and B, K_s : Public key of S, N_a, N_b : Nonce generated by A and B. Let us consider that both A and B have public key of S available.

1. $A \rightarrow S$: A requests the public key of B to server S.

2. $S \rightarrow A$: S sends public key of B, K_b to A by encrypting using its private key, A retrieves it using public key of S.

3. $B \rightarrow S$: B requests the public key of A to server S.

4. $S \rightarrow B$: S sends public key of A, K_a to B by encrypting using its private key, B retrieves it using public key of S.

5. $A \rightarrow B$: A generates nonce N_a and sends it to B by encrypting using public key of b.

6. $B \rightarrow A$: B generates nonce N_b and sends it with N_a to A by encrypting using public key of a.

7. $A \rightarrow B$: A receives nonce N_b and sends it back to B by encrypting using public key of b.

Basic motivation of this algorithm is to share a nonce with each other securely. Now let us discuss using various tools. How we can analyze security properties like authentication, secrecy, privacy, equivalence so that they are satisfied.

7.3.1   *AVISPA*

In 2006, A. Armando and D. Basic with other authors [Armando et al. (2005)] developed a tool called as a AVISPA. Full form of AVISPA is "Automated Validation of Internet Security Protocol and Applications". AVISPA is a push for the automated validation of internet security-sensitive protocols. The AVISPA tool contains AVISPA library which is a collection of 215 security protocols; among those 33 are industrial large scale protocols on which testing of AVISPA tools is done. AVISPA has detected almost all well-known attacks like man in the middle attack or forgery attack, but also at the same time has shown various new attacks for the input protocols. Functioning of AVISPA tool can be understood by analyzing following figure.

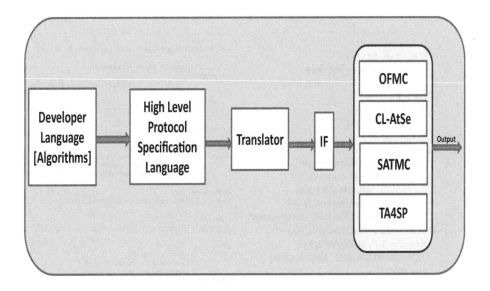

Figure 7.3: AVISPA Tool.

As shown in Figure 7.3, AVISPA tool flow can be understood as follows:

1. Developer will convert an algorithm in High Level Protocol Specification Language(HLPSL) form.

2. HLPSL2IF Translator translates a user defined security problem in to an equivalent specification written in rewrite-based formalism Intermediate Formate(IF)

3. Intermediate formate will be input to four back-end tools, On-the-Fly Model checker(OFMC), Constraint Logic based Attack Searcher(CL-AtSE), SAT

Based Model Checker(SATMC), and Tree Automata based Automatic, Approximation for the Analysis of Security Protocol(TA4SP).

4. Output will show the loop holes in protocols.

Back end tools will implement different methodology to identify state transaction for which an attack is possible on various properties of protocol.

HLPSL: High level protocol specification language is an expressive, role based, modular language that allows for the specification of control flow patterns, algebraic properties, cryptographic operators and data structures. To discuss an example for HLPSL language, we have used Needham-Schroeder Public-Key Protocol(NSPK); the following figure shows HLPSL format for the NSPK,

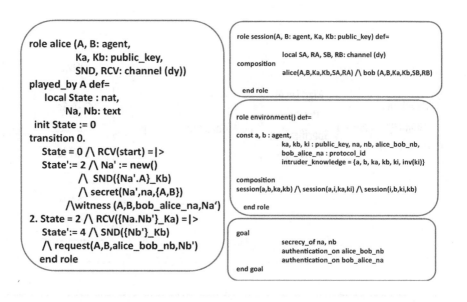

Figure 7.4: HLPSL for NSPK.

Above figure shows, first part defines role of communicating parties; we have shown the role of Alice, Each role contains the following three parts:

- Information about participant and parameters notations

- Initial state of role

- States that can be changed or called as a transitions

As shown in Figure 7.4, Alice's initial state is state 0, which means ready to receive or receive started. / symbol indicated merging of operation or parallel operations.

Transition or state change part shows the sequence of sending any parameter like nonce generating and receiving operations. Session part shows instance of each basic roles (Alice and Bob here) and composition of roles. Environment portion contains information regarding global constants, initial information about intruders, list of all agents, list of all public keys, intruder private keys and keys of other roles. Final part is goal of protocol or operations required to perform by protocols.

Basic functionalities of four back end tools can be described below:

1. **CL-AtSE:** Performs various optimization techniques and reduces or removes anomalies, redundancies, and useless breaches from protocols.

2. **OFM Checker:** Performs heuristic search operation and provides support for modeling an intruder who is capable of performing various attacks.

3. **SATMC:** Reduces input problem in to sequence of equations and builds propositional formula and input to SAT(Set of problems, Attack identifiers, Time mappers) solvers.

4. **TA4SP:** Identifies the approximation level of protocol that whether protocol is under approximate or over approximate.

7.3.2   BANLogic

BAN Logic was developed by M. Burrows [Burrows et al. (1990)]. BANLogic works based on proportional logic and how it can be used for the verification of security protocols. So BANLogic is a set of rules defined over proportional logic to analyze the information and control exchange is trustworthy and secured or not. BANLogic basically works on the belief that you can not believe on network all the data exchange is vulnerable from attackers. Basic notations followed by BANLogic are as follows:

A,B, and *S* are the principles involved and will be represented by *P,Q,R,* which ranges over these principles. K_{ab}, K_{as}, K_{bs} are the shared keys, $K_a^{-1}, K_b^{-1}, K_s^{-1}$ are secret keys, K_a, K_b, K_s are the public key of *A,B,* and *S*. All the keys, will be represented by K which ranges over encryption keys and N_a, N_b and N_c are the specific statements and *X,Y* ranges over this statement for the notations.

■ *P* **sees** *X* : *P* receives a message *X* from someone and *P* decrypts message *X* to check or to resend.

■ *P* **said** *X* : *P* sent message in past which contained statement *X*. This indicates *P* once believed on *X* that *X* is true.

■ *P* **believe** *X* : *P* trusts that statement *X* is true.

■ *P* **controls** *X* : *P* has an authority of particular statement *X* and *P* can be trusted for statement *X*. Similarly like server is trusted for keys and so server has control over keys.

■ **fresh**(X): Message X is newly generated for the current communication and same message X has not been sent earlier. This can be used for nonce, time stamp or one time used numbers.

■ $P \overset{K}{\leftrightarrow} Q$: A key K will be used by P and Q for their further communication. No other principles can get this key.

■ $\overset{K}{\mapsto} P$: K is the public key of P and its valid secret key K^{-1} is only available with P.

■ $P \overset{X}{\rightleftharpoons} Q$: Statement X is known to only P and Q. They can share with their trusted principals; it means if any other principals have X then it is shared by either P and Q. Like password is an examples of X.

■ X_K: X is encrypted using key K.

■ X_Y: With the help of Y, we can identify originator of X. So Y can be like public key.

There are five postulates that we use for the proof; let us discuss each one.

1. **Message Meaning Rule:** There are three rules; out of those two rules will be dealing with interpretation of encrypted message and one concerns the interpretation of message with keys.

   (a) Postulation of **shared keys**:

   > (P believes Q$\overset{K}{\leftrightarrow}$P, P sees X_K)/ (P believes Q said X)

   If P believes that key K is shared with Q and sees statement X is encrypted using K. Then P believes that Q once said statement X.

   (b) Postulation of **Public keys**:

   > (P believes $\overset{K}{\leftrightarrow}$Q, P sees X_{K-1})/ (P believes Q said X)

   P believes that K is shared between P and Q, P sees that X is encrypted using key K than P believes that Q also once said X.

   (c) Postulation of **secret keys**:

   > (P believes Q$\overset{Y}{\rightleftharpoons}$, P sees X_Y)/ (P believes Q said X)

   P believes that secret Y is shared with Q and sees that X is encrypted using Y then P believes that Q once said X.

2. **Nonce verification:** This is used to verify the freshness of message and it ensures that message is newly created.

   > (P believes fresh(X), P believes Q said X)/ (P believes Q believe X)

P believes that value of X is changed recently and P believes that Q once said X, then P believes that Q also believes on X.

3. **Jurisdiction Rule:** If P believes that Q is owner of X then P believes on Q on the truth of X.

$$\boxed{\text{(P believes Q controls X, P believes Q believes X)/ (P believes X)}}$$

4. **Public key - Secret Key verification rule:** If P believes that K is his public key then P must have private key K^{-1}.

$$\boxed{\text{(P believes } Q \xleftrightarrow{K} P, \text{ P sees } X_K)/ \text{ (P sees X)}}$$

If p believes that K is shared between P and Q and P sees that X is encrypted by K then P can see the statement X using K^{-1}.

$$\boxed{\text{(P believes } \xmapsto{K} P, \text{ P sees } X_K)/ \text{ (P sees X)}}$$

If P believes that K is public key of P and P sees that X is encrypted using public key K then P can see value of X using K^{-1}.

$$\boxed{\text{(P believes } \xmapsto{K} P, \text{ P sees } X_{K-1})/ \text{ (P sees X)}}$$

If P believes that K is public key of P and P sees that X is encrypted using private key K^{-1} then P can see value of X using K.

5. **Freshness rule:** If some part of formula is fresh then entire formula is fresh.

$$\boxed{\text{(P believes fresh(X)/ P believes(X,Y)}}$$

6. **Idealized rule:** Generally in any protocol, if Alice sends a message to Bob then it is written as,

$$P \rightarrow Q : \text{message}$$

but this format is often ambiguous and not suitable for formal analysis so protocol message will be idealized. For example, following protocol message,

$$A \rightarrow B: \{A, K_{ab}\}K_{bs}$$

can be idealized in the form of

$$A \rightarrow B: \{A \xleftrightarrow{K_{ab}}\}K_{bs}$$

Now let us see the BANLogic idealized form for the NSPK protocol.

- **Message 2** S → A: $\{\xmapsto{K_b} B\}K_s^{-1}$
- **Message 4** A → B: $\{N_a\}K_b$
- **Message 5** S → B: $\{\xmapsto{K_a} A\}K_s^{-1}$
- **Message 6** B → A: $\{\langle A \overset{N_b}{\rightleftharpoons} B\rangle N_a\}K_a$
- **Message 7** A → B: $\{\langle A \overset{N_a}{\rightleftharpoons} B\rangle N_b\}K_b$

After itemized statements let us analyze the protocols then we can get,

- A believes $\xmapsto{K_a} A$
- A believes $\xmapsto{K_s} S$
- S believes $\xmapsto{K_a} A$
- S believes $\xmapsto{K_s} S$
- A believes (S controls $\xmapsto{K} B$)
- A believes fresh(N_a)
- A believes $A \overset{N_a}{\rightleftharpoons} B$
- A believes fresh($\xmapsto{K_b} B$)
- B believes $\xmapsto{K_b} B$
- B believes $\xmapsto{K_s} S$
- S believes $\xmapsto{K_b} B$
- B believes (S controls $\xmapsto{K} A$)
- B believes $A \overset{N_b}{\rightleftharpoons} B$
- B believes fresh($\xmapsto{K_a} A$)

So finally we can believe the following postulates

- A believes $\xmapsto{K_b} B$
- A believes B believes $A \overset{N_b}{\rightleftharpoons} B$
- B believes $\xmapsto{K_a} A$
- B believes A believes $A \overset{N_a}{\rightleftharpoons} B$

Both A and B know public key of each other and also have knowledge about shared secrets. So A and B share nonce N_a and N_b and public key. So both A and B can securely transfer keys.

7.3.3   Scyther

[Cremers (2008)] developed scyther tool to do the analysis of security protocols. Scyther is tool for analysis falsification and automatic verification. Some of the plus points of this tool are it supports multi protocol analysis, an infinite number of set analyses, and guaranteed terminations. Scyther is a push button tool which provides a set of traces by following patter refinement algorithms.

Scyther has both GUI and command line tools for the analysis. Most probably the graphical user interface mode is used for small and medium size protocols and command line mode is used for larger and industrial protocol analysis due to its faster output. The workings of the Scyther tool can be understood by following figure,

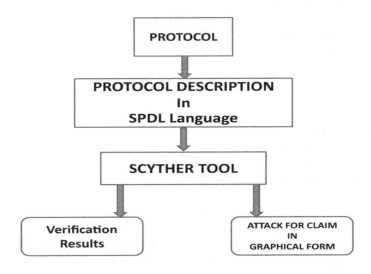

Figure 7.5: SCYTHER Tool.

As shown in Figure 7.5, the Scyther tool first provides protocol analysis in (Scyther protocol description language) SPDL language. Conversion of Needham-Schroeder public key algorithm in SPDL language is given as follow.

Needham Schroeder Public Key

  role I {

fresh Ni: Nonce;

var Nr: Nonce;

$send_1$(I,S,(I,R));

$recv_2$(S,I, pk(R), Rsk(S));

$send_3$(I,R,Ni,Ipk(R));

$recv_6$(R,I, Ni, Nrpk(I));

send$_7$(I,R, Nrpk(R));
claim$_{I1}$(I,Secret,Ni);
claim$_{I2}$(I,Secret,Nr);
claim$_{I3}$(I,Nisynch);
}

role R {
fresh Nr: Nonce;
var Ni: Nonce;
recv$_3$(I,R,Ni,Ipk(R));
send$_4$(R,S,(R,I));
recv$_5$(S,R,pk(I),Isk(S));
send$_6$(R,I,Ni,Nrpk(I));
recv$_7$(I,R,Nrpk(R));
claim$_{R1}$(R,Secret,Nr);
claim$_{R2}$(R,Secret,Ni);
claim$_{R3}$(R,Nisynch);
}

role S {
recv$_1$(I,S,(I,R));
send$_2$(S,I,pk(R),Rsk(S));
recv$_4$(R,S,(R,I));
send$_5$(S,R,pk(I),Isk(S));
}

I and R are the principles involved(Alice and Bob) and S is the server. *send* and *recv* are message exchanges, and claim indicates that some values are secret and hold with particular entities. Scyther provides attack analysis in graphical format and provides results even in case of no attack found. Scyther tool can be used in three ways.

1. Do the analysis of protocol by performing complete characterizations.

2. To automatically generates security claims for protocol and verify

3. To check whether given claims in security protocol hold or not

Scyther can be also used for analysis of certain properties like authentication and secrecy. As shown in following figure, for Needham-Schroeder protocol, there are two patterns for responder role with secret behavior of protocol and the other is MITM attack.

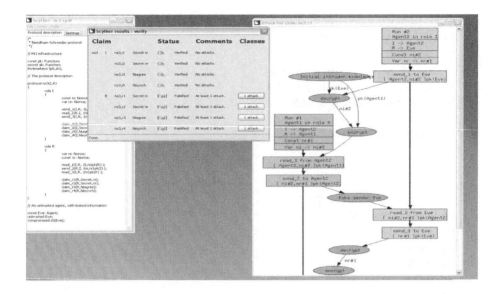

Figure 7.6: SCYTHER Tool: Needham-Schroeder Protocol.

As shown in Figure 7.6, it contains three parts; one is Scyther protocol description, an other is verification of protocol and shows one attack is there, and the third one shows a graphical representation of the attack that is claimed. Any newcomer also can easily understand the tool and how to do the analysis.

7.3.4 ProVerif

A project led by Bruno Blanchet [Blanchet (2016)] developed abstract based cryptographic protocol verifier called ProVerif. ProVerif is an automatic protocol verifier that follows Dolev-Yao Model for the modeling of an attacker. ProVerif works in maximum cryptographic operations like key exchanges, MAC calculations, hash function calculations and cypher text generation and plain text verification. ProVerif supports two types of files:

■ Horn clauses : Set of predicates combined by OR operation.

■ Pi-Calculus

With the help of ProVerif, a user can compute equivalences, authentication and secrecy. ProVerif can be installed using package manager of Objective CAML called a OPAM. Needham-Schroeder protocol can be written in ProVerif script as below:
(*
Needham-Schroeder public key protocol
Message 1: A − > S : (A, B)

Message 2: S − > A : (pkB, B) skS
Message 3: A − > B : (Na, A)pkB
Message 4: B − > S : (B, A)
Message 5: S − > B : (pkA, A)skS
Message 6: B − > A : (Na, Nb) pkA
Message 7: A − > B : (Nb)pkB

The heart of the protocol is messages 3, 6, 7.

*)

```
(* Loops if types are ignored *)
set ignoreTypes = attacker.
free c: channel.
type host.
type nonce.
type pkey.
type skey.
type spkey.
type sskey.
fun nonce to bitstring(nonce): bitstring [data,typeConverter].
(* Public key encryption *)
fun pk(skey): pkey.
fun encrypt(bitstring, pkey): bitstring.
reduc forall x: bitstring, y: skey; decrypt(encrypt(x,pk(y)),y) = x.
(* Signatures *) fun spk(sskey): spkey.
fun sign(bitstring, sskey): bitstring.
reduc forall m: bitstring, k: sskey; getmess(sign(m,k)) = m.
reduc forall m: bitstring, k: sskey; checksign(sign(m,k), spk(k)) = m.
(* Shared key encryption *) fun sencrypt(bitstring,nonce): bitstring.
reduc forall x: bitstring, y: nonce; sdecrypt(sencrypt(x,y),y) = x.
(* Secrecy assumptions *)
not attacker(new skA).
not attacker(new skB).
not attacker(new skS).
(* 2 honest host names A and B *)
free A, B: host.
(* the table host names/keys
The key table consists of pairs (host, public key) *)
table keys(host, pkey).
(* Queries *)
free secretANa, secretANb, secretBNa, secretBNb: bitstring [private].
query attacker(secretANa);
attacker(secretANb);
attacker(secretBNa);
```

```
attacker(secretBNb).
event beginBparam(host, host).
event endBparam(host, host).
event beginAparam(host, host).
event endAparam(host, host).
event beginBfull(host, host, pkey, pkey, nonce, nonce).
event endBfull(host, host, pkey, pkey, nonce, nonce).
event beginAfull(host, host, pkey, pkey, nonce, nonce).
event endAfull(host, host, pkey, pkey, nonce, nonce).

    query x: host, y: host; inj-event(endBparam(x,y)) ==> inj-event(beginBparam(x,y)).
query x1: host, x2: host, x3: pkey, x4: pkey, x5: nonce, x6: nonce;
inj-event(endBfull(x1,x2,x3,x4,x5,x6)) ==> inj-event(beginBfull(x1,x2,x3,x4,x5,x6)).
query x: host, y: host; inj-event(endAparam(x,y)) ==> inj-event(beginAparam(x,y)).
query x1: host, x2: host, x3: pkey, x4: pkey, x5: nonce, x6: nonce;
inj-event(endAfull(x1,x2,x3,x4,x5,x6)) ==> inj-event(beginAfull(x1,x2,x3,x4,x5,x6)).

    (* Role of the initiator with identity xA and secret key skxA *)

    let processInitiator(pkS: spkey, skA: skey, skB: skey) =
(* The attacker starts the initiator by choosing identity xA,
and its interlocutor xB0.
We check that xA is honest (i.e. is A or B)
and get its corresponding key.
*)
in(c, (xA: host, hostX: host));
if xA = A || xA = B then
let skxA = if xA = A then skA else skB in
let pkxA = pk(skxA) in
(* Real start of the role *)
event beginBparam(xA, hostX);
(* Message 1: Get the public key certificate for the other host *)
out(c, (xA, hostX));
(* Message 2 *)
in(c, ms: bitstring);
let (pkX: pkey, =hostX) = checksign(ms,pkS) in
(* Message 3 *)
new Na: nonce;
out(c, encrypt((Na, xA), pkX));
(* Message 6 *)
in(c, m: bitstring);
let (=Na, NX2: nonce) = decrypt(m, skxA) in
event beginBfull(xA, hostX, pkX, pkxA, Na, NX2);
(* Message 7 *)
out(c, encrypt(nonce to bitstring(NX2), pkX));
```

```
(* OK *)
if hostX = B || hostX = A then
event endAparam(xA, hostX);
event endAfull(xA, hostX, pkX, pkxA, Na, NX2);
out(c, sencrypt(secretANa, Na));
out(c, sencrypt(secretANb, NX2)).
```

(* Role of the responder with identity xB and secret key skxB *)

```
    let processResponder(pkS: spkey, skA: skey, skB: skey) =
(* The attacker starts the responder by choosing identity xB.
We check that xB is honest (i.e. is A or B). *)
in(c, xB: host);
if xB = A || xB = B then
let skxB = if xB = A then skA else skB in
let pkxB = pk(skxB) in
(* Real start of the role *)
(* Message 3 *)
in(c, m: bitstring);
let (NY: nonce, hostY: host) = decrypt(m, skxB) in
event beginAparam(hostY, xB);
(* Message 4: Get the public key certificate for the other host *)
out(c, (xB, hostY));
(* Message 5 *)
in(c,ms: bitstring);
let (pkY: pkey,=hostY) = checksign(ms,pkS) in
(* Message 6 *)
new Nb: nonce;
event beginAfull(hostY, xB, pkxB, pkY, NY, Nb);
out(c, encrypt((NY, Nb), pkY));
(* Message 7 *)
in(c, m3: bitstring);
if nonce to bitstring(Nb) = decrypt(m3, skB) then
(* OK *)
if hostY = A || hostY = B then
event endBparam(hostY, xB);
event endBfull(hostY, xB, pkxB, pkY, NY, Nb);
out(c, sencrypt(secretBNa, NY));
out(c, sencrypt(secretBNb, Nb)).
```

(* Server *)

```
    let processS(skS: sskey) =
in(c,(a: host, b: host));
get keys(=b, sb) in
```

out(c,sign((sb,b),skS)).

(* Key registration *)

let processK =
in(c, (h: host, k: pkey));
if h <> A & h <> B then insert keys(h,k).

(* Start process *)

process new skA: skey;
let pkA = pk(skA) in
out(c, pkA);
insert keys(A, pkA);
new skB: skey;
let pkB = pk(skB) in
out(c, pkB);
insert keys(B, pkB);
new skS: sskey;
let pkS = spk(skS) in
out(c, pkS);
(* Launch an unbounded number of sessions of the initiator *)
(!processInitiator(pkS, skA, skB)) |
(* Launch an unbounded number of sessions of the responder *)
(!processResponder(pkS, skA, skB)) |
(* Launch an unbounded number of sessions of the server *)
(!processS(skS)) |
(* Key registration process *)
(!processK)
)
Result of protocol from ProVerif online demo tool is as follows:
    { 200 rules inserted. The rule base contains 167 rules. 24 rules in the queue.
400 rules inserted. The rule base contains 251 rules. 15 rules in the queue.
ok, secrecy assumption verified: fact unreachable attacker skey(skA[])
ok, secrecy assumption verified: fact unreachable attacker skey(skB[])
ok, secrecy assumption verified: fact unreachable attacker sskey(skS[])
Starting query not attacker bitstring(secretANa[])
RESULT not attacker bitstring(secretANa[]) is true.
Starting query not attacker bitstring(secretANb[])
RESULT not attacker bitstring(secretANb[]) is true.
Starting query not attacker bitstring(secretBNa[])
goal reachable: attacker bitstring(secretBNa[])
RESULT not attacker bitstring(secretBNa[]) is false.
Starting query not attacker bitstring(secretBNb[])
goal reachable: attacker bitstring(secretBNb[])

RESULT not attacker bitstring(secretBNb[]) is false.
} so basic working of ProVerif is as follow:

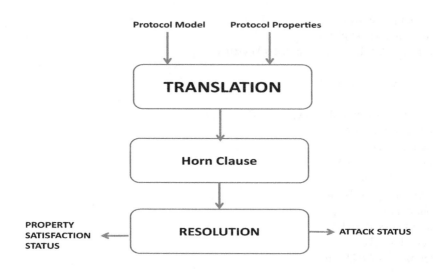

Figure 7.7: ProVerif Tool.

So as shown in Figure 7.7, ProVerif takes input as a model for cryptographic protocol (Generally PI Calculus) as shown for example protocol and security protocol that needs to be verified. It translates output into horn clauses and all combined horn clauses are considered a resolution algorithm. Horn clause represents attacker capabilities and knowledge. Basic disadvantage of ProVerif protocol is that it some times generates a false attack.

7.4 Summary

In this chapter, we have discussed the basic concept of time complexity, space complexity and energy consumption, and with the example of various cryptographic algorithm, we have shown how we can compute time consumption, space consumption and energy consumption. Later on in this chapter, we discussed the most commonly used security verification tools like AVISPA, ProVerif, BANLogic and Scyther and also discussed the example of Needham-Schroeder protocol to show the implementation of this tool.

7.5 References

Amin, R., S. K. Islam, M. K. Khan, A. Karati, D. Giri, and S. Kumari (2017). A two-factor RSA-based robust authentication system for multiserver environments. *Security and Communication Networks 2017.*

Armando, A., D. Basin, Y. Boichut, Y. Chevalier, L. Compagna, J. Cuellar, P. H. Drielsma, P. C. Heám, O. Kouchnarenko, J. Mantovani, S. Mödersheim, D. von Oheimb, M. Rusinowitch, J. Santiago, M. Turuani, L. Viganò, and L. Vigneron (2005). The avispa tool for the automated validation of internet security protocols and applications. In *Proceedings of the 17th International Conference on Computer Aided Verification*, CAV'05, Berlin, Heidelberg, pp. 281–285. Springer-Verlag.

Aslam, M. U., A. Derhab, K. Saleem, H. Abbas, M. Orgun, W. Iqbal, and B. Aslam (2017). A Survey of Authentication Schemes in Telecare Medicine Information Systems. *Journal of Medical Systems 41*(1).

Blanchet, B. (2016, October). Modeling and verifying security protocols with the applied pi calculus and proverif. *Found. Trends Priv. Secur. 1*(1-2), 1–135.

Burrows, M., M. Abadi, and R. Needham (1990, February). A logic of authentication. *ACM Trans. Comput. Syst. 8*(1), 18–36.

Cremers, C. J. (2008). The scyther tool: Verification, falsification, and analysis of security protocols. In *Proceedings of the 20th International Conference on Computer Aided Verification*, CAV '08, Berlin, Heidelberg, pp. 414–418. Springer-Verlag.

Moon, J., H. Yang, Y. Lee, and D. Won (2017). Improvement of user authentication scheme preserving uniqueness and anonymity for connected health care. *Proceedings of the 11th International Conference on Ubiquitous Information Management and Communication - IMCOM '17*, 1–8.

Nikooghadam, M., R. Jahantigh, and H. Arshad (2017). A lightweight authentication and key agreement protocol preserving user anonymity. *Multimedia Tools and Applications 76*(11), 13401–13423.

Potlapally, N. R., S. Ravi, A. Raghunathan, and N. K. Jha (2003, Aug). Analyzing the energy consumption of security protocols. In *Low Power Electronics and Design, 2003. ISLPED '03. Proceedings of the 2003 International Symposium on*, pp. 30–35.

Quan, C., J. Jung, J. Kim, Q. Sun, D. Lee, and D. Won (2017). Cryptanalysis and improvement of a biometric and smart card based remote user authentication scheme. *Proceedings of the 11th International Conference on Ubiquitous Information Management and Communication - IMCOM '17*, 1–8.

Truong, T. T., M. T. Tran, A. D. Duong, and I. Echizen (2017). Provable Identity Based User Authentication Scheme on ECC in Multi-server Environment. *Wireless Personal Communications 95*(3), 2785–2801.

Chapter 8

Future Work and Conclusions

CONTENTS

8.1   Future Work

The internet of things still has to travel a long journey in such a way so that it can develop society, change the life of people, change the method of governance, change the scenario of energy generation, change the transmission method of day-to-day amenities and change the provision of services. In future work, we are going to focus mainly on three aspects of the internet of things,

- Security aspects of internet of things communication

- Security aspects of internet of things communication

- Security aspects of internet of things communication

As you can see, all the three aspects are the same......right?

Let us try to understand basic reason behind the focus on security.

Whatever devices recently developed for the internet of things including health care devices are developed with the aim of providing services to the customer but not with the aim of providing secure services to customer. Cyber attackers are also focusing on the internet of things scenario. So in the future, communication between device to device or device to user must be secure so that except in the case of brute force attack, for the attacker there should be no chance to damage the IoT ecosystem. **Secure authentication** is the key security parameters that ensures most of the security parameters at the same time, so designing secure authentication scheme will open the door for secure IoT ecosystem. We will focus on designing authentication schemes and study various scenarios where device-device and user-device authentication happens securely.

Internet of things devices will be **resource constrained** devices so recent authentication schemes and encryption/decryption algorithms will not be suitable options for IoT devices. Rather than just secure authentication scheme, we will focus on **Lightweight Secure Authentication** schemes, in which we will focus on three major resource constraints.

- **Time complexity:** In which the reduction of complex operations will be the focus

- **Space complexity:** In which a smaller number of variables will be required to store and communicate on the link

- **Energy consumption:** In which the number of gate areas, flip flops and multiplexer requirements will be reduced

So as a future work, in a single word, we will try to design **Lightweight Security schemes** for the internet of things and its scenarios.

8.2   Conclusions

In this book, we discussed the internet of things and its security aspects. We have focused on authentication properties of security and discussed it. In this book,

- **Chapter 1:** In Chapter 1, we discussed the internet of things, internet of things architecture, internet of things security and concepts of light weight authentication.

- **Chapter 2:** In Chapter 2, we discussed mathematical fundamentals required for cryptography. We also discussed elliptic curve based operations and other security algorithms and examples.

- **Chapter 3:** In Chapter 3, we focused on IoT authentication and discussed various authentication scenarios in IoT. In this chapter, we also discussed various applications of IoT and authentication aspects of those applications.

- **Chapter 4:** In Chapter 4, we highlighted the concepts of single server based authentication schemes and various phases required.

- **Chapter 5:** In Chapter 5, we highlighted the concepts of multi server based authentication schemes and various phases required.

- ■ **Chapter 6:** In Chapter 6, we discussed various attacks and attack scenarios that need to be considered during the designing of authentication schemes; we also discussed remedies for each attack so that authentication scheme designer can try to find out about attacks in literature and design more secure schemes.

- ■ **Chapter 7:** In Chapter 7, we focused on a discussion about various cryptographic tools that are required for the analysis of security scheme. These tools will help security scheme developers to self evaluate the scheme and check the security level of the designed scheme. In this chapter, we also discussed the calculation of time complexity, space complexity and energy consumption for security protocols, so developers can easily compute the time consumption, space consumption and energy consumption and compare their own schemes with others.

Overall, in this book, we discussed all the basic fundamentals to advanced concepts required for the designing of secure authentication schemes and how to evaluate the designed authentication scheme.

Index